Basic Masonry Techniques

Created and Designed by the Editorial Staff of Ortho Books

Project Editors
Robert J. Beckstrom
Barbara Feller-Roth

Writer
Douglas Rossi

Illustrator
Tony Davis

Ortho Books

Publisher
Robert B. Loperena

Editorial Director
Christine Jordan

Managing Editor
Sally W. Smith

Acquisitions Editors
Robert J. Beckstrom
Michael D. Smith

Prepress Supervisor
Linda M. Bouchard

Sales & Marketing Manager
David C. José

Publisher's Assistant
Joni Christiansen

Graphics Coordinator
Sally J. French

Address all inquiries to:
Ortho Books
Box 5006
San Ramon, CA 94583-0906

© 1985, 1997 Monsanto Company
All rights reserved

2 3 4 5 6 7 8 9
98 99 2000 01 02

ISBN 0-89721-318-1
Library of Congress Catalog Card
Number 96-67947

THE SOLARIS GROUP
2527 Camino Ramon
San Ramon, CA 94583-0906

Editorial Coordinator
Cass Dempsey
Copyeditor
Lynne Piade
Proofreader
David Sweet
Indexer
Frances Bowles
Separations by
Color Tech Corp.
Printed in the USA by
Banta Book Group
Thanks to
John Cissel, Masonry Institute
Francis Collingwood, Ph.D.
County Line Hardware, Huntingon Sta., N.Y.
County Line Masonry Supplies, Huntington Sta., N.Y.
Deborah Cowder
William Dallwig
Jackie and Charles Davis
Robert Garbibi, P.E., National Ready Mixed Concrete Assn.
William Griffith, National Building Museum
Dave Harrison, Newsday
Tom Johnson
Dede Lange, Masonry Institute
Rose and Alessandra Lusardi
Lyngso Garden Materials
Jon Mullarky, National Ready Mixed Concrete Assn.
John Ohl, Master Gardener
Benjamin D. Rossi
Paula Rossi
Rebecca R. Rossi
Tom Shepard, Sakrete, Inc.
Brian Trimble, Brick Institute of America
David Van Ness

Photographers
Names of photographers are followed by the page numbers on which their work appears.
T=top; B=bottom; R=right; L=left
William Aplin: 48T, 78BR
Patricia J. Bruno/Positive Images: 66–67, 81T
Gay Bumgarner: Photo/Nats: 8, 18–19, 27
Richard A. Christman: 42
Alan Copeland: Back cover BR
Crandall & Crandall: 1, 4–5, 6, 33, 64, 81B, 85, 87
R. Todd Davis: 51T, 51B
David Goldberg: Front cover, 20, 47
Saxon Holt: 37, 52–53, 63B, back cover TL
Jerry Howard/Positive Images: 3B, 9, 54
Michael Landis: 40–41, 44, back cover TR
Michael McKinley: 63T, 69T, 78L, 78TR
James McNair: 69B
Ortho Photo Library: 12, 91, back cover BL
J. Parker: 3T
Portland Cement Association: 68, 88
Kenneth Rice: 62
Carol Simowitz: 48B
Jeff Stone: 38
Virginia Twinam-Smith: Photo/Nats: 58
Brian Vanden Brink: 10

Architects, Designers, and Builders
Names of architects, designers, and builders are followed by the page numbers on which their work appears.

Bob Clark: Front cover
Ron Cousino Landscaping: 47
Chad Floyd, Centerbrook Architects: 10
Tom Orchard, Boner/Orchard: 85
Rogers Gardens: 33, 64
Silverhawk & Company, Inc.: 91
Larry Steinle, ASLA, LA Studio: 6

Front Cover
Masonry, such as the brick in this inviting backyard retreat, plays a major role in the success of any landscaping project. Here, the patio, fountain, steps, and walls establish a rich framework for the plants and other landscaping features. The brick itself is as attractive as it is useful and durable.

Title Page
Elegant and timeless, a simple garden wall of stacked stone is as easy to build as it looks. All you need to know are a few techniques for choosing materials, cutting stones, and keeping the wall stable.

Page 3
Top: Covering an old patio with brick is an excellent way to transform an eyesore into a highlight.
Bottom: The techniques for working with stone are almost as timeless as the material itself.

Back Cover
Top left: Control joints, cut into fresh concrete with a grooving tool, minimize cracks in the hardened concrete.
Top right: Concrete block is a versatile material for building foundations and walls.
Bottom left: Broken concrete provides an inexpensive and attractive substitute for stone.
Bottom right: For brick walks and patios, the mortar is placed between the bricks after they have been set in a bed of mortar over a concrete base.

Basic Masonry Techniques

BUILDING WITH MASONRY

Masonry projects will enhance the beauty and usefulness of your yard for many years to come. A brick patio, for example, creates a permanent outdoor living area that is ideal for entertaining or just relaxing. A stone wall defines areas of the yard and adds beauty. A concrete or flagstone walk improves access for all seasons.

If you want to build with masonry but you're not sure how to do it, this book is for you. Easy-to-follow instructions explain the basic techniques for building walls, patios, walks, and foundations. If you can build a deck or fence, then working with brick, stone, or concrete is within your capabilities.

Before you start building, however, it is wise to develop an overall landscape plan. This chapter will get you started in the right direction. The very reason why masonry projects are so appealing—their permanence—is also the reason why you should take time to coordinate them with an overall design. Careful planning will increase your building confidence, and it can be an enjoyable process for the whole family.

In this yard, the concrete patio, brick planters, and other masonry work contribute to the overall appeal because they were carefully designed to harmonize with other elements. Their size fits the scale of the home and yard. The angles follow natural sight lines and traffic paths. The red brick provides a colorful contrast to the lush green lawn and creates a striking border for the concrete patio. The exposed aggregate finish of the concrete has interesting textures and pleasing neutral colors.

PLANNING YOUR PROJECT

What good is a new patio if your outdoor furniture doesn't fit on it or the sun never shines on it? A successful masonry project must be planned carefully to integrate with the features of your site, the needs of your family, and the principles of good design.

Developing a Landscape Design

The goal of designing a landscape is to balance the big picture with specific needs. First, consider the "givens" of your site: its size, land contours, drainage patterns, major plants, sun exposure, wind direction, noise, relationship to your home, garden structures, and so on. Also factor in practical matters such as

This stunning landscape is a minicourse in successful design. From the abstract principles of line, perspective, mass, and scale to the pleasing balance of functional needs with site realities and the skillful use of plants and masonry materials, it is a pleasure to explore.

budget, building regulations, and property values. Then balance these with how to make the space fulfill your needs. Landscape design is a process of evaluating and integrating all of these factors.

Making a Bubble Plan

Here's an easy two-step way to get your landscape design started. First, gather ideas: jot down ideas from books, clip pictures from magazines, and take photos on neighborhood walks. Keep these in a file for easy reference.

Next, draw a bubble plan (see illustration below). This will help you focus on your whole yard and keep you from getting hung up on specific details. To make the plan, first draw a freehand sketch of your property showing the fixed structures, such as the house, driveway, fences, and walkways. Include trees, sloped areas, and other natural features. Determine which

Bubble Plan

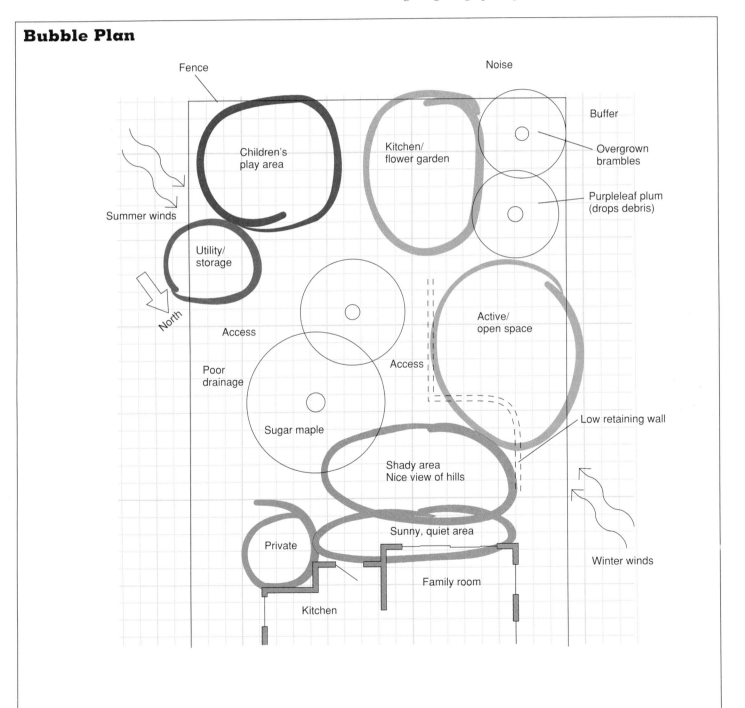

direction is north and mark it on your sketch. Also show the street and any significant features on adjacent properties. Indicate the direction of prevailing winds and the location of sunny and shaded areas and low spots that collect water. Note points of access and patterns of movement—doors, gates, and paths—including foot traffic across lawns and other open areas.

This sketch is the basis for your family to record and revise their ideas. To make full use of it, tape a piece of tracing paper over the drawing, or make several copies, and hang it where everyone in the fami-

This brick patio creates an interesting and practical framework for the trees, shrubs, and flowers planted within and around it. The amount of paved area strikes a nice balance with the planting beds.

ly can see it, such as on the refrigerator. To organize and visualize ideas, draw circles in different areas and write the suggested uses in them (this is why it's called a bubble plan). Keep a pencil nearby so that family members can add, scratch out, and change ideas.

This plan will be revised many times. You may find that the children want a play area and you want flower beds. To keep the plan from becoming unwieldy, all family members should agree on some ground rules, such as a deadline, an agreement to evaluate all suggestions, and an understand-ing that costs may be a final determinant of what can be done. Everyone in the family should participate or be represented, including future members of the family. Be sure to consider the rapidly changing needs of young children.

Considering Design Principles

Once you've decided how you want to use your landscape, you can focus on the design and shape of the actual structures. A key element in good landscaping is having a consistent style. This is where design is important. Consulting a designer knowledgeable about materials and plants, or doing research on your own, will help make your landscape project pleasing and attractive. Use the following principles to help you organize and focus your ideas.

Framework and Flowers

Landscape design has two elements: the framework, called hardscaping, consists of the walls, walkways, patios, decks, and other structures; the softscaping is the foliage. As you design your landscape, keep the two elements in mind. What kind of plantings will complement the framework, and what kind of framework enhances the plants? Visiting nurseries and garden shops, consulting with a university extension center, and studying yards and gardens that you like are inexpensive ways to research this subject.

Blending In

A rule of thumb for both hardscaping and softscaping is to choose materials that are compatible with your surroundings. Your landscape should blend in with the architectural

Stacked-stone retaining walls, made from native stone and laid in a casual, meandering pattern, blend well with this informal garden.

style of your home, the predominant landscaping materials and features of your neighborhood and community, and the native landscape. The better you understand these surroundings, the more skillfully you can blend your landscape into them and, ironically, the more successfully you can create unique and distinctive effects.

Views and Perspective

Perspective is the appearance of an object relative to its viewing point. When designing your landscape, view it from different places. You may enjoy sitting under a large shade tree at the end of the yard, looking across the landscape at the house. This is one perspective. You may look down upon your landscape from the second floor of your house—another perspective that may give you a different feeling. Consider all the perspectives, then try to devise a plan that ties together the best elements of each.

Color

This is one of the most powerful design elements you can use. A few pots of bright red flowers draw the eye and can effect perspective. Color can modify depth perception: pale plantings or surfaces help create a feeling of spaciousness, whereas dark ones help define boundaries.

When you choose masonry materials, consider their colors: dark red is a natural complement of green foliage; pink also complements, but calls more attention to itself; neutral tones, like tan and gray, are excellent background

colors; white and dark colors create stark contrast; too much of any color creates monotony.

Proportion

Patios and other surfaces should be designed with their purpose in mind. If a patio will be used for entertaining, it should be about the size of a living room, with ample space for furniture groupings. To refine the size of a patio, landscape architects often use the golden mean. Roughly translated, this rule states that for every 7 feet of house, the patio should be 5 feet. This 7:5 ratio creates a natural balance in the proportion of patio to house. As an example, a house about 28 feet wide would have a patio about 20 feet wide and 14 feet deep. A smaller patio would be acceptable, but a larger one might feel out of proportion.

Because of its modest size and clearly defined borders, this charming and intimate patio is not overwhelmed by the sweeping view.

PLANNING THE WORK

Once you and your family have decided how to use the yard space, you are ready to make a base plan and working drawings, and to devise a plan for construction. Both tasks will make your work easier later on.

Making the Base Plan

The base plan is an accurate map of the site. You'll need graph paper, tracing paper, a ⅛-inch-scale ruler, pencils, erasers, and a long tape measure (preferably 100 feet).

First, measure the perimeter of your property and the distance from the sides of your house to the property line. Then accurately draw your

Working Drawing

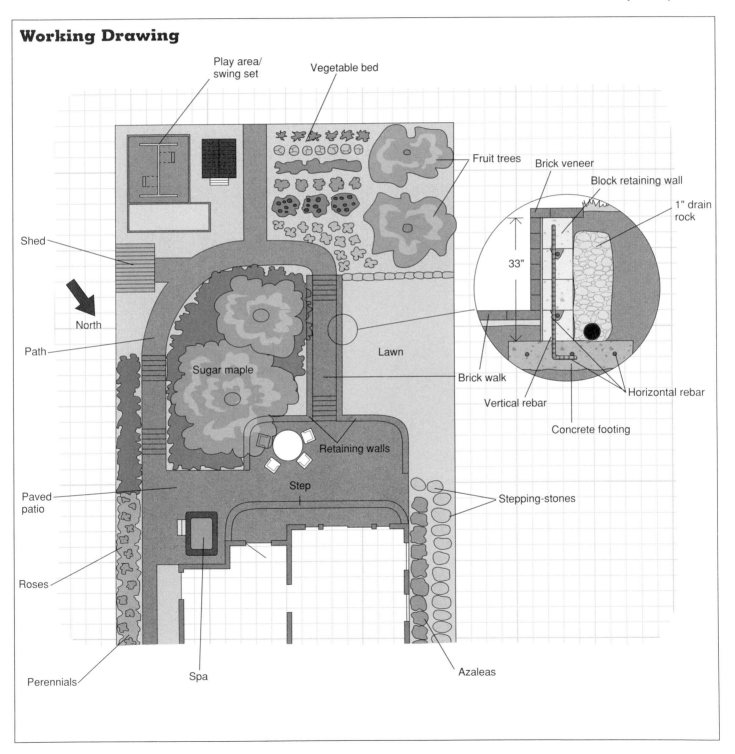

- Play area/swing set
- Vegetable bed
- Fruit trees
- Brick veneer
- Block retaining wall
- 1" drain rock
- Shed
- 33"
- North
- Path
- Lawn
- Sugar maple
- Brick walk
- Horizontal rebar
- Vertical rebar
- Concrete footing
- Retaining walls
- Step
- Stepping-stones
- Paved patio
- Roses
- Perennials
- Spa
- Azaleas

house and property lines on the graph paper. (If you have a property survey, you can copy the measurements from it.)

Now measure and add to the plan all the other significant features. These include the doors and windows of the house (for determining views and perspective); the location of exterior faucets and electrical outlets; the outlines of trees, the driveway, and paths; and the location of any underground utilities. Make sure to note the direction of north on your plan.

Making Working Drawings

With the base plan completed, you can sketch in the elements you will add: decks, patios, walks, raised flower beds, fences, electrical outlets, lighting, trees, shrubs, plants, and so on (see illustration, page 11). Do this to scale on a tracing paper overlay, because you'll probably have to change a few things around. Besides the plan view, make elevations (side views) and sections (cross sections) of any structures you

plan to build. These views show how the foundation and other details will be built.

Writing a Description of the Work

Detailed drawings help you write a description of the work, which is invaluable in developing materials lists, estimating costs, bidding the project, and scheduling the work.

Write out each step of the building process, organized by construction category. For

each category, itemize the structures that it includes. For example, if several projects require concrete work (deck footings, patio slab, steps, and so on), list all of them under Concrete Work.

If your yard is large or if you plan to phase the landscape construction over several months or even years, divide the total project into individual projects and write a work description for each. That way, you can see immediately where overlapping work, such as pouring concrete,

Some masonry work, such as handling tons of concrete in a short time, requires considerable preparation and help. Other projects, such as stacking stone or laying brick, can be done at a more leisurely pace by only one or two people.

could be done for several projects at the same time, even if you don't plan to finish all of them at once (see page 69).

Deciding Who Will Do the Work

A major decision in turning your plan into reality is deciding who will do the work. Should you hire professionals, do the work yourself, or do certain parts yourself and hire contractors for the rest? It helps to honestly assess your skills and capabilities. Are you handy with tools and knowledgeable about basic construction techniques? Can you do the work in a realistic amount of time? Do you have the time?

Landscape construction work is popular with many homeowners because such projects don't intrude on indoor living space and usually are not as demanding as remodeling the interior. There are also savings on labor costs. But also weigh the disadvantages, such as taking two months to build a patio while your family complains that the yard has become a muddy construction site. After reading this book, you will have a better idea of the scope of work and can assess your own readiness to perform it.

Hiring Professionals

If you decide to hire professionals to do some or all of the work, ask friends and neighbors for referrals. You might drive around town and see how other homes are landscaped. When you find something you like, ask the owner who designed or installed it. Many times a homeowner is happy to give a recommendation. Another source is a local garden shop, which can often recommend landscape architects, designers, or contractors.

Before selecting any professional, always ask for and check three references. Visit completed projects as well as job sites with work under construction. Ask the owners if they are satisfied with the work and with the contractor's attitude. Ask for and check bank references. Make sure that the company you select is financially stable, to protect you from liens on your property should the contractors not pay their bills.

Bidding

If you solicit bids, your drawings and written description of work become very important, because they describe what work you want done and ensure that all prospective contractors are bidding on the same work.

Don't worry if your documents are not as polished as professionally prepared construction documents. They will at least serve as the starting point for you to see how various professionals work with you at refining your ideas. This is one way of helping you decide who you're comfortable working with.

Get at least three bids. Make sure to ask if the person submitting the bid will actually be doing the work or if subcontractors will be used.

Putting It in Writing

If you decide to use a contractor, you need to have a contract to detail clearly who will be responsible for performing what work and for how much. Always insist on a well-written contract; it doesn't have to be elaborate. Most reputable contractors already have their own form. Read it carefully and don't hesitate to ask for more detail if something isn't clear. Should there be a problem in the future, the contract will be used to resolve the conflict.

A good contract will include the following (all of these may not pertain to your situation).

• Reference to construction documents: the base plan, working drawings, and a written description of the work. For small jobs a description of the work and the quality specifications of the materials will do.

• Payment schedule: usually staggered payments, with the final payment made *after final inspection.*

• Certificate of insurance from the contractor covering all risks.

• Stipulation that the contractor is responsible for obtaining permits (see right), performing the work to code, and getting necessary inspections. Make sure that startup and completion dates are specified.

When the work is completed, inspected, and approved, make sure that the contractor provides a signed receipt and lien releases before you make final payment. A lien release is a standard form stating that you have paid the contractor for the work and that you cannot be held responsible for paying suppliers or subcontractors.

Obtaining a Building Permit

When building a garage slab and some types of retaining walls, you may have to submit plans to the town building department and obtain a building permit. This may also apply if you are building a patio. Rules about when you need a permit differ from town to town. Check before construction begins.

If you do not obtain a permit for work that requires one, you run the risk of a stop-work order if the construction is discovered, or of the new construction's being discovered if you want to sell the house. Both could result in town-imposed fees or penalties, plus delays on the closing if you are selling the house.

There is another reason to inquire about the permit process. How you build a patio could mean higher property taxes. Some communities may consider a concrete patio a permanent improvement, which may result in higher taxes. The same square-footage patio built with brick or paver in sand may be taxed at a lower rate or not at all.

As part of your planning process, give the local tax assessor a call to see how your project may be taxed. You need not reveal your name or address. (If a building permit is required for your job, a copy of the permit application is usually sent to the town assessor's office. In some cases an assessor is sent out to inspect a site after the final building department approval is issued.)

MASONRY CONSTRUCTION BASICS

No matter what type of masonry materials you work with, certain preliminary tasks are common to all projects. By observing the following guidelines, you will be ready to take on any masonry project described in this book.

Getting Ready for Construction

There are several things you can do before construction begins to make the work go more smoothly. First, make sure you have the right tools. Most tasks require a few basic hand tools, such as a tape measure, hammer, level, and mason's line. Many masonry projects also require specialized tools, depending on the type of building material you are using. Some tools, such as a brickset or concrete float, are inexpensive enough that it is worth your while to buy them. Others, such as a plate compactor, rebar bender, or cement mixer, can be rented. As you read through this book, begin a list of tools you think you will need.

Construction also goes more smoothly if you have all of the materials on site before you begin work. Many are heavy, messy, or bulky, so plan in advance how you will transport and store them. Others may be easy to overlook, such as expansion joint material or masking tape to protect divider boards. Again, begin a detailed list of materials as you read through this book.

Finally, prepare yourself for construction. To some extent, any masonry project is a jour-ney into the unknown. Read through directions completely before beginning. Plan every aspect of the project as clearly as possible ahead of time.

Common Sense Safety

A backyard masonry project may not seem like a hazardous construction site, but the dangers can be similar. Observe safe work habits.

• Protect your eyes. Wear safety glasses or goggles, with side protection, whenever chipping or cutting masonry.

• Protect your back. Lift with your legs. Keep your back straight—don't twist—and bring the object up to you. Don't be shy about asking for help.

• Protect your hands and feet. Wear heavy work gloves when hammering a chisel or handling large stones. Wear construction boots, not sneakers, when doing heavy work. Don't wear loose clothing or loopy shoelaces. Use knee pads or cushions for prolonged kneeling.

• Wear a dust mask when cutting bricks or concrete with a power saw.

• Avoid fatigue. Know your limitations, especially when tempted to do "one more project" at the end of the day.

• Respect power tools. Read the manufacturer's instructions. All outside electrical outlets should have ground fault circuit interrupters (GFCIs).

Laying Out the Site

Layout lines provide an accurate reference for building straight edges, square corners, and level (or sloped) surfaces. Batter boards enable the lines to be taken down and put back up throughout the construction process. Before building the batter boards, hammer a stake where each corner of your planned project will be located. Stretch nylon mason's line from stake to stake. Using a framing square, determine roughly each right angle. Then, see how close to

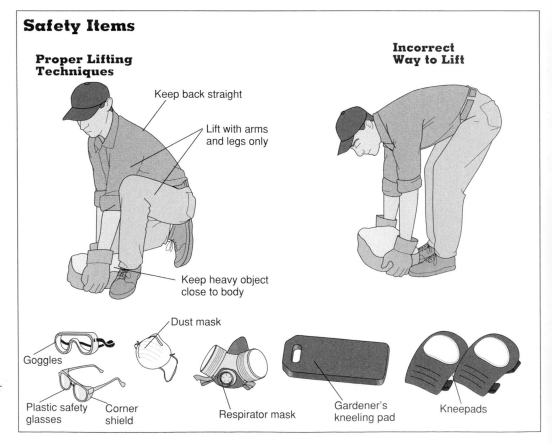

Safety Items

Proper Lifting Techniques

Keep back straight

Lift with arms and legs only

Keep heavy object close to body

Incorrect Way to Lift

Goggles

Plastic safety glasses

Corner shield

Dust mask

Respirator mask

Gardener's kneeling pad

Kneepads

Laying Out the Site

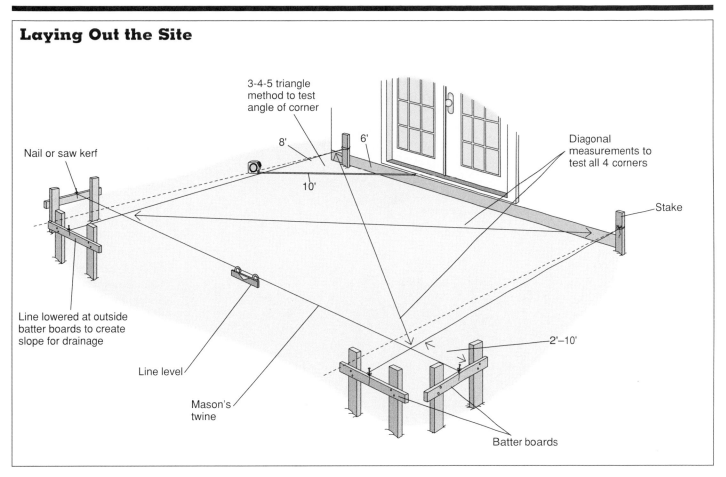

3-4-5 triangle method to test angle of corner

8'

6'

10'

Nail or saw kerf

Diagonal measurements to test all 4 corners

Stake

Line lowered at outside batter boards to create slope for drainage

Line level

Mason's twine

2'–10'

Batter boards

square the layout is by measuring the diagonals. Move the stakes until the measurements fall within an inch or two of each other.

Constructing Batter Boards

Batter boards are positioned at all four corners and consist of 2×4 stakes and 1×4 or 2×4 crosspieces. First, drive two stakes firmly into the ground for each batter board. On a small job where you will be excavating by hand, place the batter boards 4 feet behind the preliminary corner stakes. If you are going to dig ditches with a backhoe, either set the batter boards 2 feet back so that the backhoe operator can reach over them, or set them

10 feet back so that the operator can maneuver inside them.

Leveling Batter Boards

All the crosspieces should be level with each other. On small jobs, use a mason's line and line level to make level marks on all of the stakes. Then align the tops of the crosspieces with these marks, and nail or screw the crosspieces to the stakes.

For large jobs, use a water level, which is either a long length of transparent plastic tubing filled with water, or a garden hose with clear plastic attachments at both ends and filled with water. To use the water level, have a helper hold one end against one of the stakes and mark the stake at

the point where the waterline settles in the tubing. While he or she is still holding the tube against the stake, stretch the tubing to the other batter board stakes and mark each stake where the water level settles at your end of the tubing. Be sure to keep the two ends of the tubing approximately level at all times, so water does not spill out. All of the marks will be level to each other. Nail or screw the crosspieces to the stakes so the tops are flush with these marks. If the marks are too high or too low for convenient placement of the crosspieces, make new marks an equal distance up or down from the level marks.

A faster and more accurate method of leveling is to use a builder's level or a transit (see

illustration, page 16), which you can rent. These are small, tripod-mounted telescopes that remain perfectly level no matter which direction you aim them. Set the tripod far enough back from the low corner of the building site so that you can see all four corners of the site. Use the bubbles on the mounting ring to level the instrument.

To mark each stake, sight it in the eyepiece of the level or transit and have a helper mark the stake where the horizontal crosshair touches it. If the crosshair is too high to touch one or more of the stakes, have the helper hold an extended tape measure over each stake and measure down a set distance from where the crosshair

Using a Builder's Level or Transit

To use this instrument, find a spot from which you can see all batter boards clearly. Level transit, then rotate it so you can tell helper where to mark

Marking Corners

String lines

Plumb bob

Flour

Outside corner of footing trench

Temporary stake or marker

intersects the tape, then mark the stake at that distance.

Stringing the Perimeter

Mark the first side of your layout by stretching mason's line between the batter boards, aligning it over your preliminary corner stakes. Attach the string to nails driven into the tops of the crosspieces. Pull the string taut to eliminate sags.

Attach the next string at a right angle to the first, using the 3-4-5 method to establish a square corner (see illustration on page 15). Do the same on the succeeding legs.

The four string lines should create a rectangle. To check for accuracy, first measure the lengths of the strings between corners, then measure the diagonals. Adjust the strings

until the diagonal measurements are equal and each side is the required length. When you have aligned the strings, secure the nails or cut notches into the crosspieces so the strings can be taken down and then restrung again.

3-4-5 Method

The 3-4-5 method is a quick and accurate way of squaring a corner (see illustration, page 15). First, mark a point 3 feet from the corner along one string line (or house wall). Then mark a point exactly 4 feet from the corner along the intersecting string line. Now measure the distance between these points. If it is exactly 5 feet, the corner is a 90-degree angle and is perfectly square. If not, readjust the strings until the two markers are exactly 5 feet apart. For greater accuracy, use multiples of 3, 4, and 5 feet—such as 9, 12, and 15 feet.

Marking Corners

To mark the ground for excavating, suspend a plumb bob at one of the layout corners so the string barely touches the inside corner of the string lines (see illustration at left). Have a helper drive a stake into the ground exactly under the plumb bob, or mark the ground with flour or sand. Repeat this process at each corner and around the perimeter. If you are marking the ground for form boards, offset the stakes or excavation marks to allow for the thickness of the boards. The ground must be excavated far enough back so that the inside edges of the boards will align with the layout lines.

Trade Talk

Masonry has its own language. Here are some of the specialized terms that are used throughout this book.

Admixture An extra ingredient—usually in liquid form—added to concrete to change or improve a specific characteristic of the concrete, such as strength, durability, setting time, or workability.

Aggregate The rock, gravel, and sand used in making concrete.

Air entrainment A concrete admixture that creates minute air bubbles within the mix to improve its workability and the ability of the cured concrete to withstand repeated cycles of freezing and thawing.

Anchor bolt A bolt set in a concrete foundation for securing the wood mudsill to the foundation.

Ashlar A stone whose surface is squared off.

Bat Less than a whole brick. If a brick is broken in half, a bat is the part of the brick to be used.

Batter gauge A device used to measure how far the face of a masonry wall should lean back as it is being built.

Bed joint The mortar under a brick.

Brick A molded block of clay. Varied ingredients and ways of firing affect color and strength.

Bull float A T-shaped tool with a large flat blade attached with a hinged joint to a long handle. It is used to smooth freshly screeded concrete.

Cell A hollow space, or void, in a standard concrete building block.

Cinder block A hollow concrete building block made of cement and cinder, heavier than a standard concrete building block.

Closure brick The last brick in any course that must be fitted into place.

Cold joint A joint in concrete in which fresh concrete has bonded poorly to previously poured concrete.

Concrete block A generic term for a variety of hollow and solid precast blocks used in construction.

Control joint A vertical break in a concrete block wall that allows the wall sections to move up and down but still maintain lateral rigidity. A groove in a concrete slab, approximately ¼ the thickness of the slab, that controls cracking.

Course A horizontal row of bricks or concrete blocks in a wall.

Crushed rock Rock of the same type that is crushed; different from gravel.

Cure To retain moisture in concrete for a prescribed period and temperature. Curing allows the cement to chemically react with water—a process called hydration—and reach optimum strength.

Darby A long wooden float with a short handle that is used to smooth freshly screeded concrete.

Dobie A nonpermeable spacer used to keep reinforced concrete and reinforcing mesh from coming in contact with the ground.

Dry run The first course of a block wall laid without mortar, which allows for checking the layout and solving any spacing problems before making the wall permanent.

Fieldstone Stone found lying in fields or along rivers and streams.

Finishing The process of floating and troweling the surface of a concrete slab to produce the desired smoothness, density, and flatness.

Floating Finishing the surface of a freshly poured concrete slab with a hand or power float.

Gravel Small stones and stone particles ranging from ¹⁄₁₀₀ inch to ⅜ inch in size.

Grout Thin concrete mix used to fill the cells of a concrete-block wall.

Header A brick laid flat with one end facing out.

Head joint The vertical mortar joint between the ends of adjacent bricks.

Igneous rock Rock formed from volcanic action. Some igneous rock, such as granite, is very hard and makes excellent building material.

Keyway A recess or groove made in a concrete footing or wall where more concrete is poured against it later, to enable the two sections of concrete to lock together.

Laying to the line Placing block or brick about ¹⁄₁₆ inch from stretched mason's line as courses are laid.

Lead The end or corner of a wall.

Mason Building tradesperson skilled in brick, stone, or concrete construction.

Metamorphic rock Rock that has been exposed to heat and is rich in crystals and minerals.

Michigan joint A type of control joint formed by slipping a piece of roofing felt or building paper between two concrete blocks where they meet. No mortar is used on the ends of these blocks, and the paper prevents bonding when bed mortar is squeezed up.

Mortar A mixture of cement, sand, and lime used to bond bricks to each other.

Paver A manufactured stone block used like paving brick. Made of concrete, pavers come in a variety of colors and sizes, including interlocking styles.

Paving brick A dense brick made to take the weight of traffic. It is made in actual, not nominal, size because it is not used with mortar.

Rowlock header A brick laid on edge with one end facing outward.

Rowlock stretcher A brick laid on edge with the broad face exposed.

Sailor A brick stood on one end with the broad face exposed.

Screed A leveling device drawn over a freshly poured concrete slab.

Screed guide A side form that, in combination with another strip on the other side of the form, serves as a guide for striking off the surface of a concrete slab.

Shiner A brick laid flat with the wide edge facing out.

Slump A standard measurement for testing the stiffness or looseness of a fresh batch of concrete.

Soldier A brick stood on one end with the narrow edge facing out.

Stretcher A brick laid flat with the narrow edge facing outward.

Striking off Leveling the surface of a freshly poured concrete slab with a screed.

Trowel A hand tool for smoothing concrete or applying mortar and similar materials. Triangular trowels are used for brick and stone work, square trowels for smoothing concrete, stucco, and plaster.

Troweling Smoothing and compacting the surface of freshly poured concrete with a trowel.

Vibrator A power tool used to agitate fresh concrete to eliminate trapped air (but not entrained air) and settle the mix.

Wythe The thickness of a brick wall. A single-wythe wall is 1 brick thick; a double-wythe wall is 2 bricks thick.

USING BRICK & PAVERS

Bricklaying has a long tradition and some forms of the art produce breathtaking results. The basic construction technique, however, is simple: place one brick next to another. For this reason, and because bricks are uniform in size and relatively light in weight, you don't have to be an expert to achieve quick and beautiful results. Even as a novice, you can create a brick-in-sand patio in a single weekend. It might be even more fun if you involve the whole family or a few of your friends. In fact, the hardest part of your project may be choosing the brick pattern and color.

While bricks are an ancient building material, pavers are a recent product, a hybrid made out of concrete but used like brick and stone. Although not all pavers resemble bricks, the construction techniques are essentially the same as for brick. As with all the projects in this book, the keys to success with bricks and pavers are plan your project thoroughly; understand the materials and tools; take time to do the job right; and know how to fix mistakes.

Bricks can be laid in an endless variety of patterns, including circular and serpentine shapes. Broad arcs, such as the rows of bricks near the outer edge of this patio, do not require cutting the bricks, but tight, compact curves may. A well-compacted base of gravel and sand, with black poly-ethene sheeting to control weeds, ensures that this patio will remain stable for many years. For a photograph of the finished patio, see page 27.

BRICK AND PAVER BASICS

Walk into a brickyard and you will see bricks and pavers everywhere. Which one to choose? This section introduces you to the basic types, textures, sizes, and colors of these materials. You will also learn how to choose mortar, how to mix it, how to estimate bricks and mortar, how to work with them, and how to repair damaged brickwork.

All About Brick and Mortar

There are thousands of different combinations of brick sizes, colors, materials, and textures. Bricks are made at approximately 240 factories in the U.S., which usually use locally available clays. The bricks are made in kilns fired to about 1,900° F. It is the heat of the kiln and the local clay mix that give bricks their color. Certain techniques, such as higher heat and flame, add special effects to bricks.

The finish of the brick—its surface texture—depends on what process was used to make the brick. A brick with a smooth surface was probably made by the water-struck method. Sand-struck brick has a sandpaper-like surface, and wire-cut brick has a rough surface.

Types of Bricks

Besides the variations discussed above that are caused by manufacturing techniques, bricks vary according to their intended use. The types of brick most widely used are building brick and facing brick. Other categories include firebrick, paving brick, glazed brick, and a cousin of brick: concrete pavers.

Building Brick

Also called common brick or standard brick, building brick can be used in almost any brick construction project. It is generally imperfect in appearance, often arriving chipped, but in most projects the blemishes only add to the rustic quality of the brick. A brick may be solid, have holes in it (called a cored brick), or have an indentation (called a frog).

The holes and indentation help lock the brick into the mortar, and the frog should always be placed down into the mortar. Use solid bricks for walks, patios, and caps of walls. Use cored bricks where the holes will not be visible, such as in a wall or planter box.

Common bricks are divided into three grades of hardness that describe their ability to withstand the elements.

•Grade SW (severe weathering). More expensive than the others because it is made to withstand the harshest weather. It is recommended for projects in areas with sub-zero winters, or in any project where the brick will be in contact with the ground, such as retaining walls and patios.

•Grade MW (moderate weathering). Used in areas that have subfreezing weather but not likely to be permeated with water. Use this grade for above ground walls, but not for patios.

•Grade NW (no weathering). Not designed to withstand any severe weather without risk of cracking. It is a good choice for many interior projects but not for outdoor projects.

Facing Brick

Facing brick is used where appearance is critical. It usually has a nicely finished surface and is strong and durable. In projects where there are several wythes (layers) of brick, the interior wythes are usually common building brick while the exterior brick is facing brick.

Facing brick comes in three different grades.

•FBX. For walls requiring near perfect units.

•FBS. For general interior/exterior work where some variation in color and surface is acceptable.

This sampler of bricks and pavers includes: Back row, left to right, firebrick and tan and gray concrete pavers; center row, two types of red common brick, three types of manufactured used brick, and brick with variegated colors; front row, firebrick, red brick, and used brick pavers.

• FBA. Used for variety in color and texture and for architectural effects.

Firebrick

Easily identified by its yellow color, firebrick is made with a special clay and fired at extremely high temperatures to harden it. This brick is used as a lining for fireplaces or barbecues because of its heat-resistant qualities. When using firebrick, you must use a special refractory cement or fire-clay mortar that won't fall apart under high temperatures.

Paving Brick

Harder than common brick, paving brick is full sized (not nominal) for use *without* mortar and is used for brick-in-sand patios or driveways.

Paving brick is classified by how much traffic load it can bear and its ability to resist weathering. For the homeowner, use type 3 or type 2 for load capacity. For weathering, use class SX for harsh wet winters; MS for exterior use in mild climates; and NX for interior work.

Pavers

These are not true bricks—they are concrete—but they are used like paving brick. Pavers, which are more durable than bricks and less expensive, come in many shapes, sizes, textures, and colors. Pavers are used for walks, driveways, patios, and walls. Some manufacturers make their pavers so they interlock, assuring a solid surface.

Other Types of Bricks

There are five other types of bricks for residential projects: used bricks, manufactured used bricks, brick veneers, glazed bricks, and clinkers.

Used bricks often have a warm and rustic quality. However, used brick is generally expensive, because you are paying for the labor that went into removing the old mortar. If you can haul used bricks away from a demolition site and clean them yourself, you can save a considerable amount of money. One caution about used bricks: they may not be as weather resistant as new brick or may be of low quality, particularly if they are thirty or more years old. Modern bricks use better clays and improved firing

techniques. An alternative to used bricks is manufactured used bricks—they are building bricks made with gray and white splotches to look like used bricks.

Brick veneers are the same face size as standard bricks, but only 1 inch to ½ inch thick. These are usually used to cover interior walls.

Glazed bricks are face bricks that were treated with liquid glazing before being fired in the kiln. The surface is glassy. This kind of brick can be used in kitchens and bathrooms because it is easy to clean.

Clinkers are rough, hard, and can be dark or deformed bricks that are the result of high heat in the kiln. They are usually used for decorative purposes.

Brick Terminology

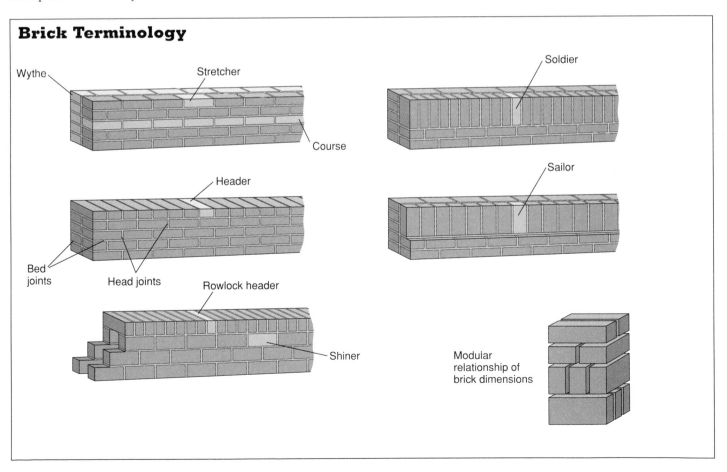

Wythe · Stretcher · Course · Soldier · Header · Sailor · Bed joints · Head joints · Rowlock header · Shiner · Modular relationship of brick dimensions

Brick Sizes and Colors

Bricks come in many sizes. Like lumber, bricks are referred to by their nominal dimensions rather than exact sizes (except for paver bricks). Some have a nominal length of 8 inches and a width of 4 inches, but are actually 7⅝ inches long and 3⅝ inches wide to allow space for a ⅜-inch-thick joint. Unless you are doing exceptionally precise work, you needn't worry about this difference. The recommended joint thickness in this book is ⅜ inch.

Bricks are made in "modular" sizes. This simply means that bricks are sized in increments of 4 inches, so that they will fit together regardless of how they are placed. Bricks are commonly 4 inches wide and 8 inches long, so that one brick will fit across two bricks laid side by side. Modular bricks make it simpler to fit walls and patios together.

Virtually all brick colors fall in the range of red, brown, or yellow, but the variations can be tremendous. Except for high-quality facing brick, most brick is not uniform in color and these irregularities add to the charm of brick in the landscape.

In addition to the varied colors available, bricks come in different textures. The more common textures include smooth face, ruggs face, stippled face, and matte face. When considering bricks for a patio or sidewalk, choose a texture that will provide a nonslip walking surface. Smooth-faced bricks tend to be slippery and, unlike con-crete pavers, tend to grow mildew in humid climates.

Mortar

Mortar is a combination of portland cement, hydrated lime, sand, and water. Portland cement binds together all the elements of the mortar compound. Hydrated lime makes the mortar more plastic and easier to work. Sand adds volume to the mortar.

You can buy premixed mortar, or order bulk materials and mix your own. Buy premixed mortar for small jobs; it is more expensive but saves time. One 60-pound sack will be enough for about 40 bricks with ⅜-inch-wide joints.

For large jobs, it is best to buy all the ingredients separately and mix your own, since proportions will vary depending on the type of brickwork you are doing. Buy masonry cement, which is mixed with lime, and sand. Sand delivered in bulk will be wet, so adjust the amount of water in your mix accordingly. The sand should be clean, so store it on sheets of plastic or plywood.

Mortar Grades

Different types of mortar mixes are available. It is important to select the right

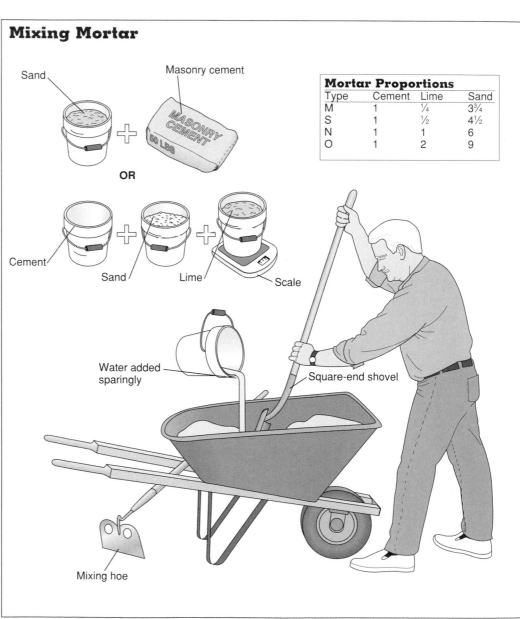

Mixing Mortar

Mortar Proportions			
Type	Cement	Lime	Sand
M	1	¼	3¾
S	1	½	4½
N	1	1	6
O	1	2	9

Sand
Masonry cement
MASONRY CEMENT
50 LBS

OR

Cement
Sand
Lime
Scale

Water added sparingly
Square-end shovel
Mixing hoe

type depending on the project, location, and climate.

- Type M. A high-strength mortar suitable for general use and in load-bearing walls. Use this mix for any masonry that comes into contact with the ground, like foundations or retaining walls.
- Type S. This is not as strong a mortar mix as Type M. Recommended for above-grade exterior use.
- Type N. This is a medium-strength mortar for use above grade where high compression or lateral strength is not required, such as in freestanding brick walls.
- Type 0. A low-strength mortar that can be used for interior brickwork or where the bricks will be exposed to little weathering and no freezing.

For the projects described in this book, Type N is a good, all-around mortar mix that is strong enough for most yard projects. However, if you are building in an area with subzero temperatures, use Type M.

Mixing Mortar

There is nothing difficult about mixing mortar; just be careful to use the correct proportions, measured by weight. For precise measurements, use one bucket for each ingredient, set it on a bathroom scale to weigh the ingredient for the first batch, and mark the bucket for measuring identical amounts of the ingredient for the rest of the batches.

Mortar is most easily mixed in a contractor's wheelbarrow that can then be readily moved about the site as you work. Place the measured amounts of dry ingredients in the wheelbarrow and mix them together thoroughly. Then push the ingredients into the middle to make a hill. Scoop out the hill and slowly pour the water into the depression. Gradually drag the dry ingredients into the water with a hoe or a flat-bottomed shovel—not a pointed shovel. Use a watering can or bucket to add the water and not the hose, otherwise you may add too much. Be especially careful, when adding the last amounts of water, not to add too much.

A good way to test mortar consistency is to trowel it up into a series of ridges. If the ridges appear dry and crumbly, add water; if the mortar ridges slump immediately, it is too wet. In a proper mix, the ridges will stay sharp and firm. Adjust the mix as needed.

Mortar will become too hard to work with after about two hours, so mix only as much as you can use in that time period. If you see that the mortar is getting a little stiff, you can add some water to return it to a good working consistency. Try not to do this more than twice with one batch; too much water weakens the mortar.

Estimating Brick and Mortar

To estimate brick and mortar for your project, first calculate the square footage of wall or patio surface area. For a quick estimate of bricks, figure five bricks per square foot of patio surface, and 7 bricks per square foot of wall surface (for 4-inch wide walls). For a more accurate estimate, consult the estimating chart below. It shows how many bricks and how much mortar you will need for every 100 square feet of surface area, depending on the type of project, size of brick, and thickness of joints you are planning.

To figure the area of a square or rectangle, multiply length by width. All dimensions should be in feet, or fractions of a foot—not inches. For a circular patio, multiply 3.14 by the square of the radius. For more complex shapes, first draw them on graph paper, with each square on the graph paper equal to 1 square foot. Then count all the squares that are more than half inside the border of the patio or walk to determine the square footage.

After you have figured the total number of bricks for your project, add 5 percent for breakage. Make additional allowances for features of your project where the brick pattern differs from the field, such as borders around patios, or a wall cap of bricks set on edge. For instance, if you were going to cap a 20-foot wall with 4-inch-wide bricks laid side by side, you would need an additional 3 bricks to cap each foot of wall, or 60 bricks.

Estimating Brick and Mortar

Materials Required Per 100 Sq Ft Surface Area

Size of Brick (in inches)	Thickness of Wall	Number of Bricks ⅜" Joint	Number of Bricks ½" Joint	Amount of Mortar	Mortar Ingredients (mixed 1:3 by volume) Masonry Cement	Mortar Ingredients (mixed 1:3 by volume) Sand
For Brick Walls:						
2¼ × 3⅝ × 7⅝	4"	675	–	8.1 cu ft	2.7 sacks	8.1 cu ft
	8"	1350	–	19.4 cu ft	6.5 sacks	19.4 cu ft
2¼ × 3¾ × 8	4"	655	–	9.0 cu ft	3.0 sacks	9.0 cu ft
		–	616	11.7 cu ft	4.0 sacks	11.7 cu ft
	8"	1310	–	21.0 cu ft	7.0 sacks	21.0 cu ft
		–	1232	27.5 cu ft	9.1 sacks	27.5 cu ft
For Brick Paving:						
2¼ × 3⅝ × 7⅝		450	–	6.75 cu ft*	2.25 sacks	6.8 cu ft
2¼ × 3¾ × 8		–	400	7.4 cu ft*	2.5 sacks	7.4 cu ft
1⅝ × 4 × 8		450 (no mortar joints)		(no mortar required)		

*Includes ½" bed under bricks

Brick Tools

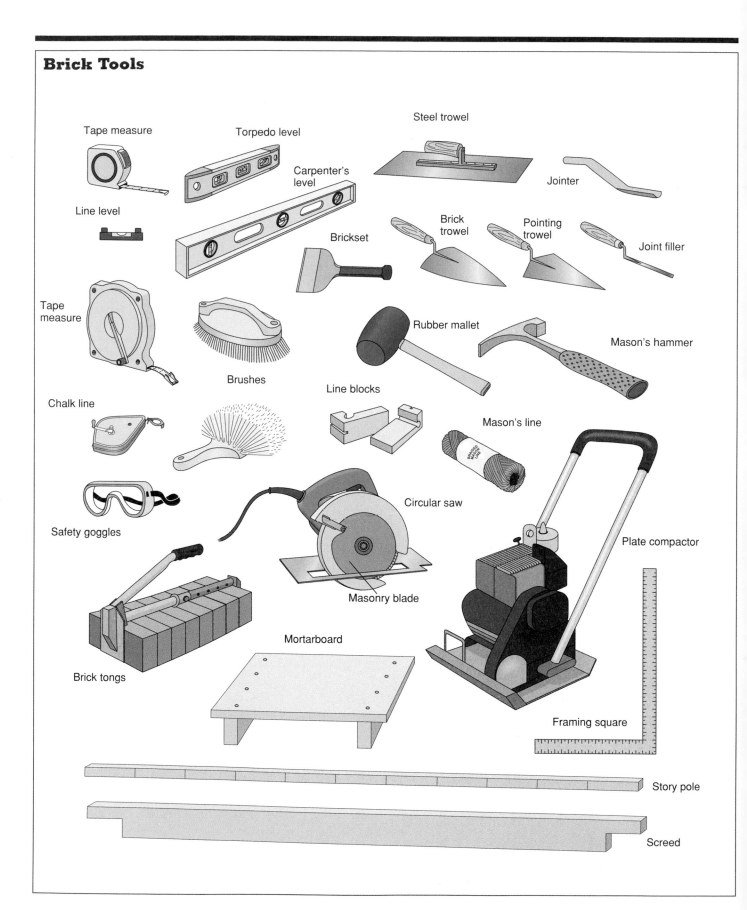

Tape measure

Torpedo level

Steel trowel

Carpenter's level

Jointer

Line level

Brickset

Brick trowel

Pointing trowel

Joint filler

Tape measure

Rubber mallet

Mason's hammer

Brushes

Line blocks

Chalk line

Mason's line

Safety goggles

Circular saw

Plate compactor

Masonry blade

Brick tongs

Mortarboard

Framing square

Story pole

Screed

Working with Brick

Brick is easy to work with. All you need are the right tools and knowledge of a few basic techniques. This won't make you a master mason, but you will have a foundation for learning bricklaying skills. The tools and cutting techniques are the same for any project and are presented here. Installation techniques depend on the project (see pages 27 to 39).

The Toolbox

Although there is a wide variety of tools available for brickwork, you can handle many different projects with only a few of the basic ones. The tools listed alphabetically below will help you decide what you need, depending on your particular project.

•Brick hammer, or mason's hammer. One end of head is for hammering, the other for cutting and shaping brick. Handle, made of wood, fiberglass, or steel with a rubber grip, must be able to absorb shock.

•Brickset. Wide chisel for splitting and dressing bricks.

•Brick tongs. Used to clamp and carry up to 10 bricks at a time.

•Brush, with soft bristles. For sweeping sand into fresh masonry joints.

•Brush, with stiff bristles. For cleaning bricks and dried masonry joints.

•Chalk line or chalk box. For marking straight lines.

•Chisel. Mason's chisel, with a beveled cutting edge, used to score and cut brick.

•Framing square, or carpenter's square. For measuring right angles.

•Jointers, in various shapes and sizes. Used for smoothing and finishing the mortar joints between bricks. A short length of ½-inch copper pipe can be used for convex joints.

•Level. Spirit level, at least 4 feet long, for keeping work level and aligned. (Wood levels are easier to clean.)

•Line blocks. Wood or plastic blocks with grooves on the inner edges to hold mason line for building walls.

•Mason line. Thin nylon line that, unlike ordinary string, can be pulled taut and held in place without sagging. Used to establish straight reference line.

•Masonry blade. An abrasive blade for circular saws, for cutting or scoring bricks.

•Masonry saw. Special saw that can cut bricks and which can be rented.

•Mortarboard. Board for holding a shovelful or two of mortar, to be scooped up with the trowel. Can be a 2-foot-square piece of plywood with two 2×4s nailed to the bottom to raise it off the ground, or a handheld aluminum board.

•Plate compactor. A rental power tool that packs the gravel base, or firms up the bricks on sand, by vibration.

•Pointing trowel. A small trowel for pushing mortar into joints, which are usually ⅜ inch or ½ inch wide, and for smoothing the mortar.

•Rules. A 6-foot folding rule or retractable steel tape with numbered guides that are used to check the height of brick and mortar joints. In addition, you should have a 25-foot steel tape measure for laying out patios, walks, or walls (and a 100-foot tape measure for large projects).

•Safety glasses or goggles. For all chipping, cutting, and sawing of brick.

•Story pole. A straight 2×2 for holding against brickwork to check alignment. Mark the height of each course of bricks and mortar joints on it.

•Trowels. One triangular brick trowel with a 5½-inch-by 10-inch blade, and one smaller trowel for detail work.

Cutting Brick

If you have ever watched a professional bricklayer at work, cutting a brick looks simple. Many pros just hold the brick with one hand and strike it with the edge of their trowel, and zap!—a cleanly cut brick. In the real world, it's not always that easy.

A convenient way to cut a brick is with a brickset. Wear safety goggles for all cutting and sawing. First score the brick along the cutting line on all four sides with the chisel end of a mason's hammer or the brickset. Then set the brick on a firm and level surface, and place the brickset on the scored line, facing the bevel

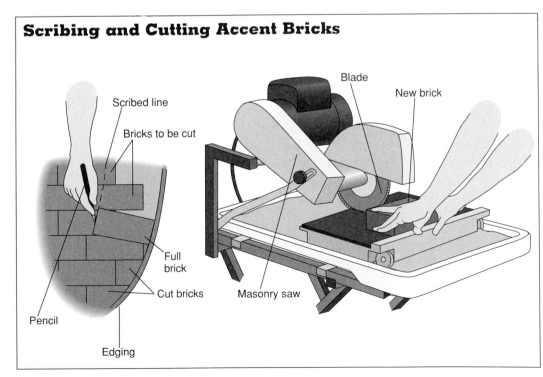

Scribing and Cutting Accent Bricks

Scribed line

Bricks to be cut

Blade

New brick

Full brick

Cut bricks

Masonry saw

Pencil

Edging

toward the waste section of the brick. Then strike the brickset with a single forceful blow of a hammer. Use the chisel end of the mason's hammer to smooth any ragged edges.

Using a brickset is less than satisfactory for cutting perfectly straight edges or for cutting angles (especially for patios with a diagonal brick pattern that requires numerous angled cuts along the edges). It is also virtually impossible to cut hardened bricks with a brickset. The solution is to use a power saw with a masonry blade. Buy one for your circular saw, or rent a masonry saw and blade.

It is important to secure the brick before cutting it with a circular saw. Use a vise or a folding worktable with a top that clamps. Place the brick in the clamp, tighten it up, and cut. (Using C clamps won't work because they get in the way of the saw.) If, while you cut, the blade does not go completely through the brick, turn the brick over and cut the other side. Don't try to sever it with a hammer blow; it will leave a jagged edge.

To cut a brick with a masonry saw, which you can rent, set the brick on the table and align the cutting mark with the saw blade. Start the saw, then gently lower the blade onto the brick. Move the blade back and forth over the brick with a minimum of downward pressure—let the blade do the work. Keep your fingers away from the saw blade. If the saw is the type with a stationary blade and movable table, slide the table slowly toward the blade until it cuts through the brick.

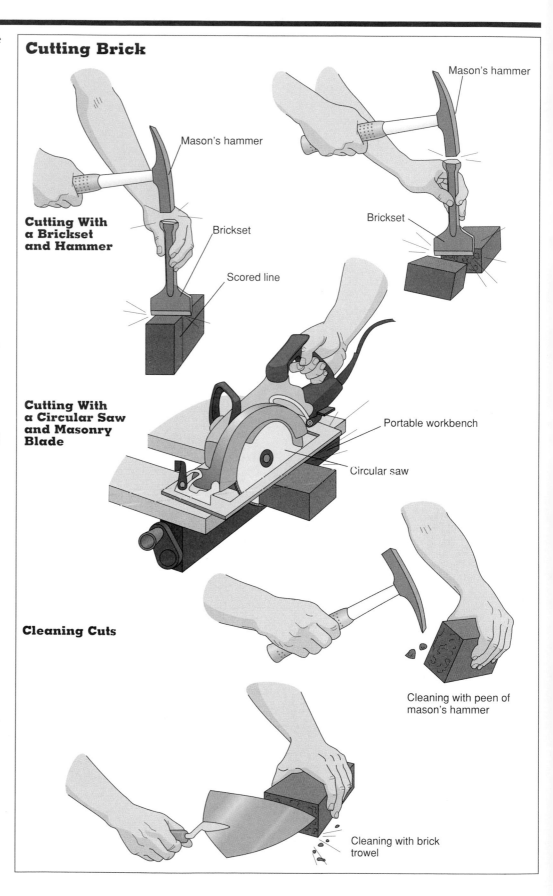

Cutting Brick

Cutting With a Brickset and Hammer

Mason's hammer

Mason's hammer

Brickset

Scored line

Brickset

Cutting With a Circular Saw and Masonry Blade

Portable workbench

Circular saw

Cleaning Cuts

Cleaning with peen of mason's hammer

Cleaning with brick trowel

Building with bricks is an ideal backyard project. The bricks are easy to handle, the techniques for installing bricks are simple, and the project can be built in stages over several weekends—just continue laying the bricks where you left off.

Building a Brick-in-Sand Patio or Walk

Laying bricks or pavers in a smooth bed of sand is one of the easiest and fastest ways to install a patio, a walk, or even a driveway. The keys to success with this technique

are a well-compacted base and a firm edging around the perimeter. The bricks are laid tightly together within the edging. Sand is swept over the bricks; it falls between the cracks and forms tiny wedges that keep the bricks in place.

Planning the Project

Before going to a brickyard to look at samples, first decide on the pattern you'd like to use to lay the brick. Although you may ultimately select one of the brick paving patterns shown on page 28, before deciding on one, sit down to view your landscape and think about the effect you wish to achieve. Remember that your efforts will last a long time, so site your project where it will be most appropriate and most pleasing to you.

For example, the area beneath a large tree where shade

inhibits lawn growth might be the perfect spot for a brick patio. Then, instead of a standard brick pattern set in a rectangle, think about a circular patio with a curving brick pattern that emphasizes the swirls of the tree. Or you could build a walk that meanders in an arc around the tree.

Of all the brick patterns illustrated, only the diagonal herringbone requires extensive and careful cutting. Renting a masonry saw will simplify that chore.

After you have chosen the bricks and the pattern, you may want to adjust the overall

This semicircular patio, also shown on pages 18 and 19, has been installed over a bed of compacted gravel and sand. The border of "soldier" bricks around the edge holds the outer bricks in place.

Brick Paving Patterns

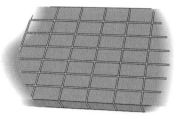

Jack-on-jack

Running bond

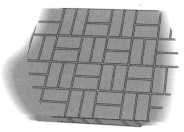

Basket weave

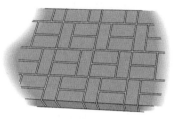

Half-basket weave

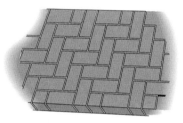

Herringbone, 90°

Herringbone, 45°

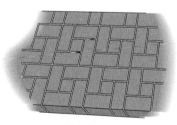

Pinwheel

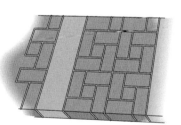

Pinwheel with concrete dividers

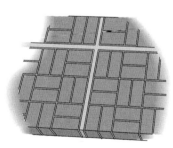

Grid pattern

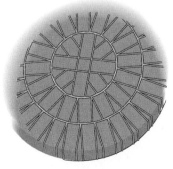

Whorled

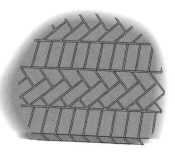

Herringbone and soldiers

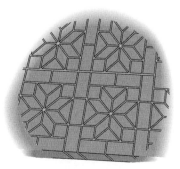

Mediterranean

dimensions of your project so that the bricks come out evenly and you don't have to cut too many bricks. For more information about planning the dimensions, see pages 10 and 84.

Establishing the Layout

For layout techniques, refer to the first chapter (see page 14) and to the techniques for laying out a concrete patio (see page 80). If your patio abuts the house or one side is parallel with a house wall, begin the layout along that wall and square all of the corners to it. Remember to build slope into the layout so that the patio will drain away from the house.

To lay out a walk, stretch two parallel strings between stakes driven into the ground or, for a curved walk, position two hoses on the ground, parallel to each other, curving in the desired shape. Mark the dimensions on the ground with chalk or flour, then remove the strings or hoses.

Providing for Drainage

When planning a patio or walk, don't forget to consider potential drainage problems. If the patio or walk will be near the end of a downspout on the side of your house, extend the downspout with underground drainpipe to carry water away from the house. Dig a trench across the proposed patio or walk and connect the downspout to a flexible drainpipe placed in the trench.

If there is no natural drainage area for the water, you can build a dry well and direct the water into it. To construct a dry well, dig a hole 4 feet wide and 4 feet deep, then fill it with rock or gravel. Extend the drainpipe to the dry well, cover the pipe with 3 to 4 inches of gravel, and backfill with topsoil.

A patio next to a house should slope away from the house so that water runs off it. The excavated base, the gravel bed, and the brick surface should all be sloped, at a rate of 1 inch per 10 feet of distance. A walk should be crowned so that it slopes downward toward both edges.

Installing Edging

Since edging prevents shifting in a brick-in-sand patio or walk, it is essential that the edging be sturdy and long lasting. The five common types are rot-resistant wood, bricks on end, concrete, stone, and plastic (see illustration below). Concrete edging can either be flush with the tops of the bricks or lowered by the thickness of a brick and then finished with brick edging set in the wet concrete.

Whenever possible, make a dry run with the bricks in your chosen pattern to see where the edging should be placed. This will minimize the number of bricks that must be cut to fit inside the edging. On a large patio or driveway, you can install permanent edging on two adjacent sides and temporary edging on the two opposite sides. Lay bricks to the temporary edging and adjust it until it fits against your pattern, then replace it with the permanent edging.

If the patio is to be raised above the surrounding grade, a good choice for edging is 4×4, 6×6, or 8×8 treated landscaping timbers. Set them with half their thickness in the ground and fill the interior area with sand to within the thickness of a brick from the top. Butt the ends of the timbers and toenail them together with galvanized 5-inch nails. To keep them in place, drill

Edgings for Patios and Walks

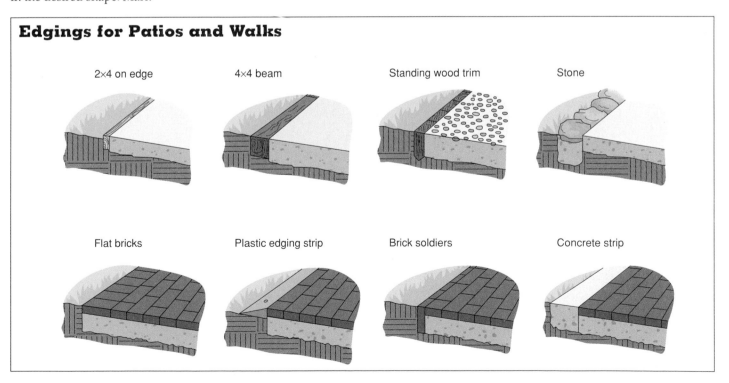

2×4 on edge   4×4 beam   Standing wood trim   Stone

Flat bricks   Plastic edging strip   Brick soldiers   Concrete strip

⅜-inch holes near both ends and in the center of each timber, then drive 3-foot lengths of ½-inch concrete reinforcing rods (called rebar) through the holes into the ground.

Edging that is flush with the ground can be made quickly and inexpensively from 2×4s of redwood, cedar, cypress, or pressure-treated wood. To install edging boards, dig a trench approximately 5 inches deep under the perimeter string lines. Place the boards with the inside edge directly under the string lines. Support the boards with 12-inch 1×2 stakes, driven every 4 feet and nailed to the outside of the boards. Make sure that the tops of the stakes are an inch or so below the top of the edging boards.

Measure down from the string lines to keep the edging boards even. If a board is not even, don't try to remove the stake; just pry it up or pound it down until the board is level. When the edging boards are completed, cut off the tops of the stakes at a 45-degree angle and backfill the outside of the trench.

An alternative to board edging is a row of bricks placed on end around the perimeter. Bricks in this position should be set in the ground so that their tops will be flush with the surface of the patio.

Dig the edging trench so that the inner face of the bricks is directly under the string line. Use a level to keep the tops of the bricks level. Setting the bricks in concrete makes this edging more permanent.

Grading the Site

With the edging in place, excavate and level the area to a depth of at least 4 inches (see page 80). Use a straight 2×4 with a carpenter's level on top to check your work. If the ground is soft and loose, it should be tamped down before adding sand. You can make a tamping tool from a 4-foot length of 4×4 with two handles nailed to the top, or rent a plate compactor.

If the ground is wet for much of the year, which may cause the completed brickwork to settle unevenly, excavate the ground deep enough for at least 2 inches of gravel under the sand. Tamp down the gravel with a hand tamper, or rent a power compactor. The compactor is heavy, so make sure that you've got a helper to move it to the site. Once you start the compactor, it may feel as though it's taking off, but get control of it and make several passes over the gravel. Go back and fill in any depressions and compact the area again.

Tip: Order gravel and sand and have them delivered. To estimate how many cubic feet you will need, multiply the height (0.2 for 2 inches) times the width times the length, then divide by 27. Order unwashed coarse sand; it is cheaper than washed fine sand and works well for a base.

Building the Sand Base

Using a shovel, fill the excavated patio or walk area with clean, pebble-free sand to within one brick depth of the final surface. Dampen the

sand, then compact it to provide a firm base for the bricks.

The next step is to screed the sand base so it is just the thickness of a brick below the top of the edging. Make a screed board by either notching the ends of a 2×6 or nailing ears on the ends of a 2×4 (see illustration opposite).

Place the screed board on the edging and draw it across the sand. If the patio is too wide for a screed board to reach across it, use a temporary support made from a board staked and leveled down the center of the patio site.

A large patio or walk could be divided into sections with redwood or pressure-treated boards that will be permanent. Nail these permanent dividers to the perimeter edging and support them in the center with sand.

Before arranging the bricks or pavers, place a layer of weed-blocking fabric over the sand. The fabric allows water to penetrate down through it but inhibits weeds from growing up through it. Roofing felt will also work, but you have to punch in a lot of holes for drainage, through which weeds can grow. Using fabric also prevents ridges of sand from forming between the bricks as you place them.

Placing the Bricks or Pavers

Starting at one corner, arrange the bricks or pavers in your selected pattern. Place them firmly on the weed-blocking fabric and snugly against one another. Avoid sliding them into place. Then rap them lightly with a hammer handle

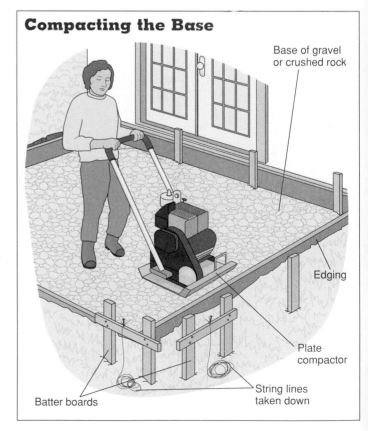

Compacting the Base

Base of gravel or crushed rock

Edging

Plate compactor

String lines taken down

Batter boards

Building a Brick-on-Sand Walk

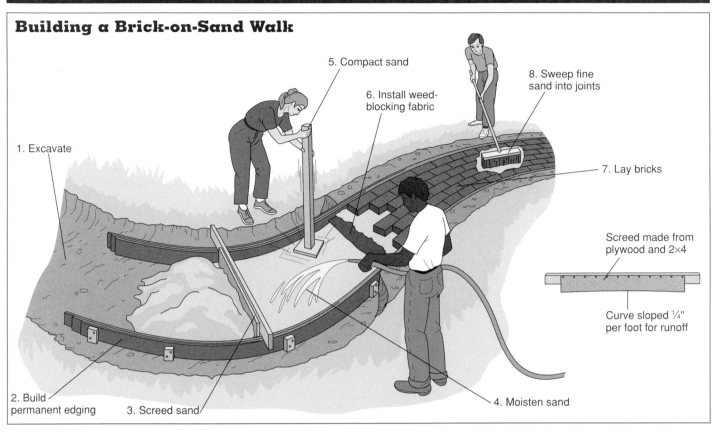

5. Compact sand

6. Install weed-blocking fabric

8. Sweep fine sand into joints

1. Excavate

7. Lay bricks

Screed made from plywood and 2×4

Curve sloped ¼" per foot for runoff

2. Build permanent edging

3. Screed sand

4. Moisten sand

Building a Brick-on-Sand Patio

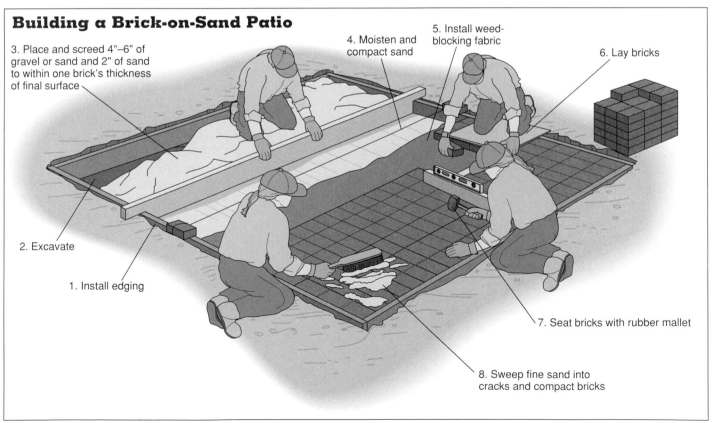

3. Place and screed 4"–6" of gravel or sand and 2" of sand to within one brick's thickness of final surface

4. Moisten and compact sand

5. Install weed-blocking fabric

6. Lay bricks

2. Excavate

1. Install edging

7. Seat bricks with rubber mallet

8. Sweep fine sand into cracks and compact bricks

or rubber mallet to settle them, or just place them and go over them later with the power compactor.

As you continue placing the bricks, use a straight 2✕4 to keep the joints aligned, or run a string from edging to edging and use it to keep the brick courses straight.

To distribute your weight and keep from pushing individual bricks or pavers out of place, kneel on a half sheet of plywood or a few boards placed on top of the bricks you've already laid.

After seating the bricks with a mallet or the compactor, sprinkle a thin layer of fine, dry sand over the bricks, then sweep the sand back and forth so that the grains start to work their way into the joints between the bricks. This step is important, because the grains act as tiny wedges to keep the bricks from moving. (When hosing off the patio or walk later, don't hose out the sand.) Settle the bricks and sand with a plate compactor, or leave the sand on the patio for several days; continue to walk on and sweep the patio until the sand works its way fully into the joints.

Making Accent Cuts

Cutting bricks or pavers for curves and other accents is the touch that will make your project look professional.

The easiest way is to wait until the patio or walk is completed except for the bricks or pavers adjacent to the edging that need cutting. Then scribe all the cuts as described below, rent a power masonry

saw (bring a helper for loading it), and cut all the accent bricks at once.

To scribe and cut bricks (or pavers), hold a full brick against the edging and move it around the perimeter of the work. Wherever the brick you're holding overlaps a full brick you've already set, mark the overlap on the set brick with a pencil.

Next, moving around the work, pick up each marked brick and replace it with a full brick. Cut the marked brick at the pencil line with the power masonry saw. The cut portion that was not overlapped will now fit the gap between the full brick and the edging.

Setting Bricks in Mortar

A permanent brick patio or walk requires setting the bricks in mortar over a concrete base.

The concrete base, or slab, should be at least 4 inches thick and have steel reinforcement to minimize cracks. In areas with severe winters or unstable ground, it should have a gravel base and, if required by local conditions, be 6 inches thick. For more information about building a concrete patio, see pages 80 to 83.

After the concrete has hardened, install a temporary edging or 2-by lumber around the perimeter to serve as a screed guide. The top of the boards should extend above the concrete surface to accommodate the thickness of one brick, plus ½ inch of mortar. At this time you should wet the bricks (see opposite).

Next, starting at one end, spread mortar over a section of the slab and screed it smooth. The screed board should have a notch at each

end that is the same depth as the thickness of one brick. Then lay bricks in the mortar in your chosen pattern, using a string line or straightedge to keep the courses aligned. Place ⅜- or ½-inch spacers between the bricks to hold them apart for the mortar joints. Check your work by placing a level across the bricks. Repeat this process until the slab is covered.

After 24 hours, or when the mortar has set, fill the joints between the bricks with mortar. Carefully place it with a brick trowel, scraping residue from the bricks as you go. After the mortar is firm enough to accept a thumbprint, smooth the joints with a jointing tool. Wipe off the remaining residue with a damp sponge. Wait for the mortar to set, then make a final pass with a stiff brush.

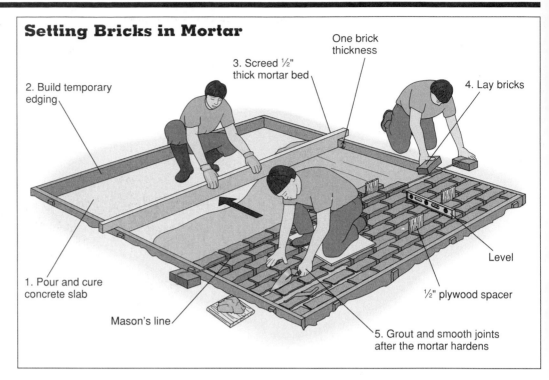

Setting Bricks in Mortar

2. Build temporary edging

3. Screed ½" thick mortar bed

One brick thickness

4. Lay bricks

1. Pour and cure concrete slab

Mason's line

Level

½" plywood spacer

5. Grout and smooth joints after the mortar hardens

BUILDING A BRICK WALL

Laying bricks looks easy when it's done by a professional. You may not be able to work as quickly, or produce a perfectly straight surface, but if you take your time and work carefully, you can build a beautiful brick wall.

Planning the Wall

Before you begin the wall, consider the following.

• Location, height, length, and thickness (number of wythes).

• Brick type and color, and brick-laying pattern.

• Mortar type and color (if desired).

• Depth of footing and amount of concrete (see page 58).

• Delivery and storage of brick and materials. When digging your footing trench, don't throw the dirt where it will get in the way of brick and materials deliveries. Have the materials delivered and positioned before going to the next step.

Using Wet or Dry Bricks

Some bricks, when laid in fresh mortar, absorb water from the mortar and weaken the joint. Other bricks do not. Because of the vast variety of bricks, there is no easy answer to the question, should I wet the bricks? Here is a simple test you should perform on your bricks before laying them.

Select any brick at random and pencil a circle on its face about the size of a quarter. Pour ½ teaspoon of water onto the circle, then time the rate of absorption. If the water is absorbed into the brick in 90 seconds or less, the bricks should be wetted before laying them.

The easiest way to wet bricks is to spread out several dozen at a time and spray them on all sides with a hose. Let the surfaces dry before using them, however, because a wet surface will not bond with mortar.

Bricklaying Techniques

If you have never worked with bricks and mortar before, it is a good idea to practice first before actually starting a project. Handling a brick, mortar, and a trowel at the same time will seem awkward at first, but because bricklaying is very repetitive, you will soon feel comfortable at your work (see page 34).

Mortar is not *placed* with a trowel, it is thrown—kind of like picking up a fried egg with a spatula and slinging it onto a plate (it's all in the wrist). To practice this technique, first place a shovel full of mortar on a mortarboard. Cut a section from the edge of the mound of mortar and, with the trowel, shape the section of mortar roughly the length and width of the trowel. Pick up the mortar by sliding the edge of the trowel under it in one quick motion. As you pick up the mortar, snap the trowel slightly to bond the mortar to the trowel. This amount of mortar should be enough to bed at least 3 bricks.

Now comes the part that takes practice. The mortar should be thrown from the trowel in a sweeping motion, with the mortar sliding off the edge of the trowel and landing firmly on top of the bricks. You want to avoid placing the mortar by dribbling it off little by little. Not only is this much slower, but the mortar does not bond to the bricks as well. Start by trying to cover 2 bricks with one throw, then 3 or 4 bricks.

Spread the mortar to an even thickness of about 1 inch, then use the trowel to cut off

A continuous concrete footing below ground keeps this brick wall from settling and cracking. A cap that combines whole and half bricks is wide enough to cover the double-wythe wall and create a generous overhang.

How to Lay Brick

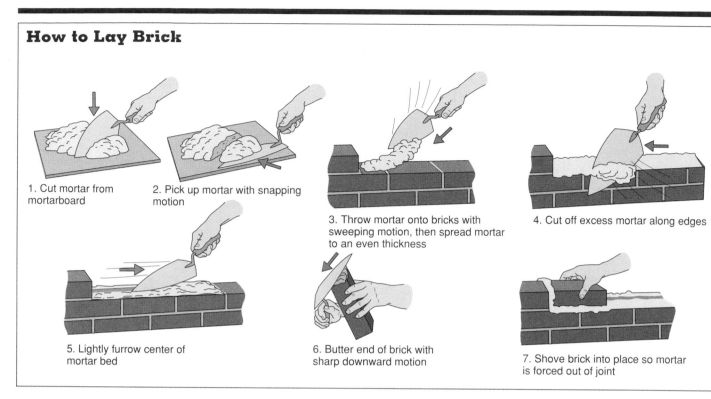

1. Cut mortar from mortarboard

2. Pick up mortar with snapping motion

3. Throw mortar onto bricks with sweeping motion, then spread mortar to an even thickness

4. Cut off excess mortar along edges

5. Lightly furrow center of mortar bed

6. Butter end of brick with sharp downward motion

7. Shove brick into place so mortar is forced out of joint

the excess mortar along the edges. You can add that mortar to the bed or place it back on the mortarboard.

Once the mortar is spread on the wall, lightly furrow the center of the bed with the tip of the trowel. This allows the mortar to adjust to an even thickness as the brick is placed on it. Avoid making a deep furrow, which may leave an air gap under the brick.

After the first brick has been placed, the end of the next brick to be placed must be "buttered." Hold the brick in one hand, with the end to be buttered tipped up at a 45-degree angle. Place a small amount of mortar on the trowel and slap it on the end of the brick with a sharp downward motion. This technique causes the mortar to bond to the brick.

A brick is not just put in place; it must be shoved. After buttering the end of the brick, place it on the mortar bed and shove the brick firmly against the one already in place. When shoved correctly, mortar will be forced out the sides and top of the joint. Skim off that excess with the trowel and use it to butter the end of the next brick. Rap the placed brick with the end of the trowel handle to set and level it.

If you set a brick too low, lift it out, add new mortar, and lay it again. Simply pulling it up slightly would leave a gap where water could enter, freeze, and crack the mortar.

Building a Single-Wythe Wall

One of the easiest types of brick walls to build is a single-wythe brick planter box (see opposite). You wouldn't normally build a single-wythe

wall more than a foot or so high, because it could be knocked down easily. However, once you put corners on such a wall, it becomes much stronger.

A few brick planter boxes filled with bright flowers will enhance the yard or garden and also provide you with practice for bigger projects.

Laying Out the Wall

This three-sided rectangular planter box uses the concrete foundation of the house to form the back wall. Build the footing for the planter box (see page 44), then lay out the exact position of the wall on the footing.

First brush the footing clean of any dirt so that the mortar will bond tightly to the footing. Center a brick on the footing next to the foundation and make a pencil mark where the outside edge of the first brick

will be. Lay a steel framing square flat on the ground, with one edge against the foundation and the tongue next to the pencil mark. Now, using the square as a guide, snap a chalk line on the footing.

Use the square again to mark all the other corners, first marking where the outside edge of the brick will be, then aligning the chalk line with the square and snapping the line. To double-check that your layout is square, measure the diagonals. If your work is accurate, the measurements will be the same.

The framing square is accurate enough for a wall that will be 10 feet or less in length. For larger projects, use what is called the 3-4-5 method (see pages 15 and 16) to make sure that your layout is square.

Building a Brick Planter

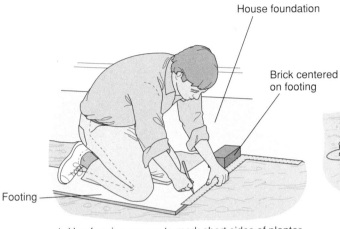

House foundation

Brick centered on footing

Footing

1. Use framing square to mark short sides of planter

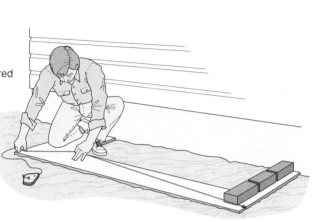

2. Mark long side of planter with chalk line

3. Lay first course dry with ⅜" spacers

4. Spread mortar and shove first brick into place. Keep mortar off guidelines

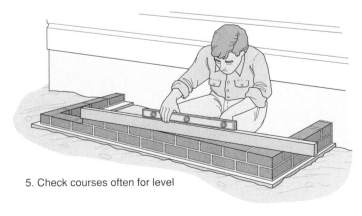

5. Check courses often for level

6. Tool joints—vertical (head) joints first, then horizontal (bed) joints

Laying the Bricks

With the wall dimensions outlined on the footing, lay the first course of bricks in a dry run (this is called dry bonding) to see how they fit. Space them with pieces of ⅜-inch plywood, or wedge the tip of your little finger tightly between each brick, which will space them almost exactly the right amount. If the footing was laid out correctly, the bricks should fit. If they don't, adjust the lines on the footing so that you don't have to cut a lot of bricks.

Using the troweling technique described on pages 33 and 34, lay a bed of mortar on top of 4 bricks. Try not to cover the chalk line. Make a shallow furrow in the center of the mortar. Butter the end of the first brick, then shove it against the foundation and into the mortar bed.

Lay the rest of the bricks in this first course in the same manner, then place a level on top to see if any bricks are out of line. If a brick is too high, tap it down with the trowel handle; if one is too low, remove it, add more mortar, and lay the brick again.

Use the trowel to cut off excess mortar on both sides of this first course, then use the tip of the trowel to smooth and firm up the mortar on the footing.

At the first corner, spread mortar on the footing for 3 or 4 bricks; butter the end of the first brick and shove it into place against the side of the last brick. Continue laying the bricks down this second side of the planter. Since this leg may be longer than your 4-foot level, place a straight

2×4 on top of the bricks and place the level on top of it to check your work. For longer walls, use mason's line and line blocks (see illustration).

When starting the second course of brick, begin with half a brick, so that this course will offset the first course. The corners of each course should be offset.

Keep working your way around the planter box, course by course, until you reach the desired height of 1 to 2 feet. Remember to tool the joints before they get too hard (see below).

Tooling and Finishing

It is important to keep a careful watch on the mortar, and tool the joints before the mortar becomes too hard. Joints should be tooled when the mortar will just accept a thumbprint with firm pressure. The tools used are called jointers.

Tooling the joints compresses the mortar. This is necessary to completely fill the spaces and keep out any moisture. Always tool the head (vertical) joints first, then the bed (horizontal) joints. If there is not enough mortar in the joint, place a small amount of mortar on the trowel, then pick up some of it with the jointer and press it into the joint.

The type of joint you use will affect both the appearance and the lasting quality of the wall. At right are eight common types of joint.

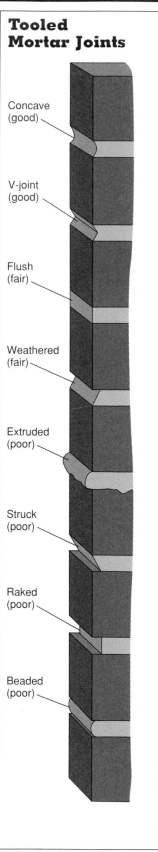

Tooled Mortar Joints

Concave (good)

V-joint (good)

Flush (fair)

Weathered (fair)

Extruded (poor)

Struck (poor)

Raked (poor)

Beaded (poor)

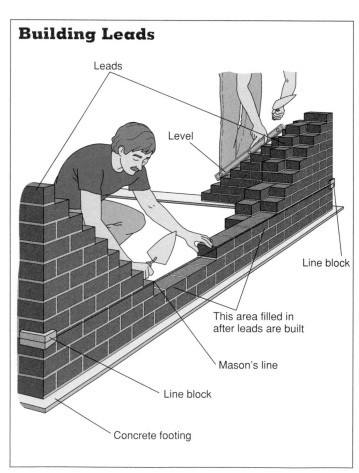

Building Leads

Leads

Level

Line block

This area filled in after leads are built

Mason's line

Line block

Concrete footing

•Concave joint. This is one of the most common types of finish and one of the most weatherproof.

•V joint. Like the concave joint, this type is highly effective at preventing water from entering the joint.

•Flush joint. The mortar squeezed out between the bricks is simply scraped off flush with the face of the wall. This method is fast but not recommended for use in areas with heavy rain or freezing temperatures.

•Weathered joint. Here the tip of the trowel is run along the base of the upper brick to cut it back ¼ inch at a 45-degree upward angle. This joint resists weathering fairly well, but it is structurally weaker and less attractive than either the concave or V joint.

•Extruded joint. The mortar that is forced out between the bricks is simply left as is. The joint is subject to weathering and should not be used in areas with heavy rain or freezing temperatures.

•Struck joint. This is the opposite of the weathered joint, with the trowel used to compact the mortar along the upper edge of the joint. It is weak because it allows water to collect in the lower edge of the joint.

•Raked joint. For dramatic shadow lines on a wall, the raked joint is best. A special metal raking tool is used to remove all the mortar up to ½ inch deep between the joints, or a piece of hardwood is cut to fit between the joints. Because raked joints allow water to collect in them, they should not be used in areas with freezing temperatures.

•Beaded joint. This is the opposite of the raked joint. A special metal tool is used to leave a shaped bead of mortar protruding from the joint.

After the joints are tooled, use a soft brush to sweep away loose particles of mortar. Use a large wet sponge to clean mortar off the bricks before it dries and stains them.

Capping a Wall

Walls are commonly capped for both a finished appearance and protection. Walls can be capped with bricks or with contrasting material such as pavers or flagstone.

A common way to cap a wall is with bricks placed as headers (bricks laid flat with one end facing out) or rowlock headers (bricks laid on edge with one end facing out) (see illustration, page 21). To do this, lay a bed of mortar on the top course. Then, depending which style you are using, butter either the edge or the face of each brick and position it. Because this will be a particularly visible part of the wall, watch carefully that the joint thickness remains constant.

When you are within about 4 feet of the end of the wall, dry-bond the remaining cap bricks to see how they will fit. *Plan your cuts!* Since the last brick may extend over the end of the wall, it will have to be cut for a flush fit. Don't place this cut brick last on the cap. Instead, place it 3 to 4 feet from the end, where it will not be noticed.

If you cap the wall with flagstone or pavers, use the same principles described above for brick.

Although the mortar joints may be tooled in any pattern you wish, the joints along the cap should be struck with a flat jointer so that there is no depression to collect moisture.

Building Leads

For walls higher than the brick planter box described above, the corners of the wall, called leads, are built first, then the bricks are laid on the line between them.

The reason that leads are built first is so that mason's line can be stretched the length of the wall to keep each course straight and level.

There are two commonly used types of lead: the straight lead—for the ends of walls—and the corner lead. The wall shown at left is a basic single-wythe corner lead with a running bond pattern.

In laying out the corner lead, remember that the number of bricks in the first course equals the number of courses in the finished corner. For example, if your wall will be 11 courses high, you will lay out 5 bricks down one leg of the corner and 6 down the other leg.

Once the lead is built, you will see that it forms stair steps by half a brick at a time. To make sure that all the bricks are properly positioned, place a level or straightedge

The brick caps on these gate pillars show how the modular dimensions of brick can be put to effective use. Three bricks laid on edge equal one full brick length, a ratio that makes it possible to keep the caps symmetrical.

against the lead so that it runs from top to bottom. The corners of the bricks should just touch the straightedge. If a brick is out of line, tap it gently into place, but don't move the top or bottom brick.

Since you are laying bricks from the leads toward the center, the last brick laid in each course must be fitted between two previously laid bricks. This is called the closure brick. To properly lay this brick, butter the ends of both bricks already laid, then butter both ends of the closure brick. Place the closure brick directly over the opening and force it down into position. A considerable amount of mortar will be squeezed out, but this is necessary to form a tight seal.

Installing Drainage

There are two basic drainage systems, depending on the amount of water expected to collect behind a wall. One system carries water away from behind the wall; the other lets water drain through the wall (see illustration).

The first type of drain is excellent for a retaining wall—one that is built next to a bank or along sloping ground—which may be subject to considerable pressure when the ground becomes saturated with heavy rain or snowmelt. To prevent the wall from being pushed out of line or actually toppled, good drainage behind the wall is essential.

The preferred method is to place rigid, perforated drainpipe—the type used for septic field leach lines—in a bed of gravel behind the wall. The pipe should be positioned with the holes facing down—not

up, as you would think. This allows the water to percolate up into the pipe and then run down the narrow channel between the two rows of holes.

For brick planter boxes, or any wall holding soil in place, an effective drain can be made simply by leaving a ⅜-inch gap (the width of the mortar joint) between the bricks in the base of the wall.

Alternatively, you could insert lengths of ¾-inch plastic pipe through the base of the wall. Cut about 1 inch off the end of a brick every 4 feet to accommodate the pipe. Extend the pipe about 6 inches into the soil behind the wall and surround it with gravel. Drill holes along the sides of the pipe to allow more water to enter the pipe.

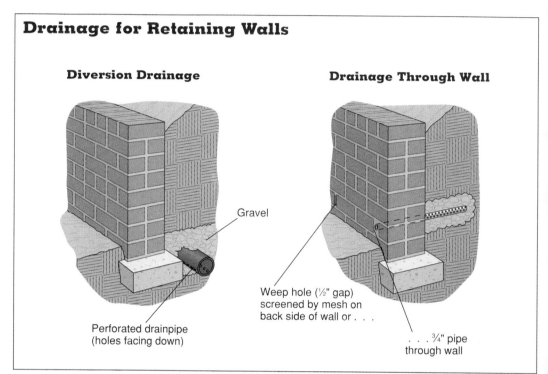

Drainage for Retaining Walls

Diversion Drainage

Drainage Through Wall

Gravel

Perforated drainpipe (holes facing down)

Weep hole (½" gap) screened by mesh on back side of wall or . . .

. . . ¾" pipe through wall

An abundance of brick—in the patios, planters, and retaining walls—is held in check by an even greater abundance of greenery. The overall effect is rich texture and an interesting interplay of angles and shapes.

Brick Problems and Repairs

Efflorescence

Efflorescence is a white powder on the surface of brick; occasionally it is a green stain. It is caused by water pressure forcing soluble salts from the brick and mortar to the surface.

If water pressure originates from the soil beneath a walk or behind a wall, drains should be installed to prevent future moisture problems before the brick is cleaned. If it is caused by water that soaked into the brick because of poor mortar joints, a cleaning or two may do.

To remove efflorescence, first scrub it thoroughly with a burlap bag or a stiff fiber brush. Do not use a metal brush, because particles of metal will cause rust stains. Then use high water pressure to wash the brick.

If the problem persists, wash the brick with muriatic acid—1 part acid added to 10 parts water. Follow label instructions carefully. First, thoroughly soak the brick with plain water. Scrub the affected area with the acid solution, then hose it off thoroughly. Because the acid may change the color of the brick slightly, scrub the entire area, or at least wash the surrounding area with a weaker solution to blend it.

Cracked Mortar Joints

The causes of joint cracks range from freezing to settling to old age. The old mortar must be removed and replaced with new mortar, a process known as repointing or tuckpointing.

Remove cracked and loose mortar with a hammer and a cold chisel with a tapered blade. Wear protective eye goggles. Cut the old mortar out of the joints to a uniform depth of ½ inch to 1 inch. Use high-pressure water to remove all particles of mortar in the joints; brush the joints clean, then dampen them before repointing.

Mortar for repointing is different from that used to lay brick. Make the mix from 1 part portland cement, 2 parts hydrated lime, and 8 parts sand. Add only enough water initially to make a ball of mortar. Let this mix stand for about 20 minutes, then add enough water to make the mixture pliable and plastic. This rehydration technique reduces the tendency of old mortar to draw too much water out of the new mix and weaken it. Even so, spray the old brick and mortar with water a few minutes before you start repointing.

To repoint, place some mortar on a mortarboard or trowel, hold the board next to the joint, and push mortar into the joint with a pointing tool. Fill the horizontal joints first, then the vertical joints. Smooth all the joints with a jointer, starting with the vertical joints, and finishing with the horizontal joints.

Graffiti

Off-the-shelf products to clean painted graffiti off brick are available, but not all cleaning products are the same and some may actually make things worse. Call a local landmarks preservation organization for a list of recommended products and cleaning methods.

It is important to follow the manufacturer's directions and to test the product in an inconspicuous area. You probably won't remove the paint completely and will still see a paint "ghost."

Cracked, Broken, or Loose Brick

If a mortared brick is broken or loose it must be chiseled out (see illustration). Use a narrow-bladed cold chisel to remove the mortar and brick. Wear goggles. Break the brick with the chisel, if necessary, to speed up the work.

Carefully chip away all the old mortar on the surrounding bricks, then sweep or vacuum out all debris. Thoroughly wet the new and surrounding bricks, then let them dry; otherwise the mortar will not bond properly.

Mix a small batch of mortar that is most suitable for your climate (see page 22). Butter the sides of the replacement brick. Carefully center it in the opening, then press it into place. Excess mortar will be forced out, but this is necessary to ensure a tight joint. Clean the edges of the brick with a trowel. When the mortar can accept a thumbprint with firm pressure, tool the joints.

Repointing

1. Remove old mortar

2. Wet joints and fill with new mortar

3. Tool joints

Replacing a Damaged Brick

Brickset or mason's chisel

1. Remove damaged brick

Pry bar

2. Line cavity with fresh mortar and insert new brick

USING BLOCK

Concrete block is more than just, well, concrete block. This badly maligned building material of industrial, institutional, and utility construction also has its place in the home landscape. It can be used for a wall to separate parts of your yard, as a backdrop to a perennial border, to direct the eye to other areas, or as a boundary between properties. It is especially suited for retaining walls. Made from many different kinds of materials and coming in numerous shapes, sizes, and colors, it is appropriate for decorative as well as structural purposes. It is easy to build with, which makes it an attractive option for residential use. Even easier to use are the interlocking types of block, which require no mortar. And it can be covered by veneer finishes such as stone, brick, tile, or stucco with excellent results.

This chapter gives you the basics of building with concrete block, including several projects that will add a professional-looking finish to your landscape.

Using concrete block for foundation walls avoids having to build elaborate concrete forms. It also eliminates the need for delivering loads of ready-mixed concrete, an advantage in remote areas that are not readily accessible to heavy cement trucks.

BLOCK BUILDING BASICS

Building almost anything out of block is an easy and satisfying do-it-yourself project, as long as you aren't building walls over 3 feet high. You can choose blocks from a wide range of sizes, types, colors, and textures, or cover them with interesting surface finishes.

Types and Sizes of Block

Once you determine how you want to use block, the next step is to visit several masonry supply yards.

One factor that will determine your choice is whether you are building your wall on a concrete footing or not.

Walls built of standard building block require concrete footings. You may opt instead for interlocking block, some of which does not require a concrete footing. Instead, you lay block directly on gravel in a shallow trench. The first course of block is buried (see page 50).

Standard Building Block

This concrete block is the type with two or three hollow spaces, which are called cells. The dividers are called webs. The webs and face edges are wider on one side of the block than the other. The wider side should always face up and be buttered with mortar before another block is placed on it.

Standard building block is nominally 8 inches wide, 8 inches high, and 16 inches long. Block widths of 4, 6, 10, and 12 inches are also available, as is 8-inch-square block, called half block. All these dimensions are actually ⅜ inch

less to allow for a ⅜-inch mortar joint. Each block weighs 40 to 50 pounds.

The most commonly used blocks for wall construction are stretcher, corner, and half block (see illustration opposite). Stretcher block has ribs at each end, which are buttered and fitted together. Corner block has a ribbed end that is butted to the adjacent block, and a smooth end that is exposed. Half block is commonly used at the ends of walls; half the length of a standard block, it allows the joints to be staggered. When you estimate the amount of block for your project, don't forget to figure in the placement and amount of corner blocks and half blocks that you'll need.

Block is made in three classes: solid load-bearing, hollow load-bearing, and hollow non-load-bearing. The projects described in this chapter use the hollow load-bearing type.

Specialty and Decorative Block

In addition to standard building block, you should be aware of special block, such as block with curved ends for a decorative touch on a wall, or block with a solid top used for capping a wall.

Other variations include block with split faces that look like hand-hewn stone, block with sculptured faces, and slump block, which closely resembles adobe brick.

Mortar

The same mortar mixes described on pages 22 and 23 for brick are used for block. The mortar should be mixed as for

Just one corner of a well-stocked masonry yard suggests the wide range of sizes, shapes, textures, and configurations of concrete blocks that is available.

brickwork but with a little less water to make it stiffer. If it is too wet it will ooze out of the joint due to the weight of the block. Mix only as much mortar as you can use in about an hour.

Planning the Project

Planning your project should begin with a rough sketch; this will minimize the amount of cut block in your layout. If you're careful you can use full and half block and avoid cutting any to an odd size. The way to do this is to plan the project so that the horizontal and vertical runs are an even number of feet plus 8 inches, or an odd number of feet plus 4 inches.

Because concrete block is made with constant dimensions—more so than brick and certainly more than stone—you can plan quite precisely how much block you will need for any given project. Just measure the length of the run, divide by 16 (the length, in inches, of a standard block), and multiply by the number of courses. Add about 10 percent for waste (see illustration above).

If your project contains openings that require block of odd lengths, you can cut the block to fit with a mason's hammer and chisel, or you can make neater cuts with a power masonry saw and masonry blade, which can be rented.

Concrete block does not offer the great variety of bonds (mortar joints) and patterns that brick does. For almost all projects, the standard running bond is best.

This is not only the strongest bond, it is simple and straightforward as well.

Unlike brick, concrete block should not be wetted before use. The block should be covered if rain or a heavy morning dew is expected.

For other techniques for working with block, see the projects described below.

Tools for Block Work

Concrete block work requires relatively few tools—about the same as those used for brickwork (see pages 24 and 25). These include a trowel, 4-foot level, mason's line, line blocks (to keep the courses straight), a mortarboard (to keep the concrete mix close to your work), and jointers (to smooth and finish the joints). You can purchase a convex or V-shaped jointer, or make a convex jointer from a 22-inch length of ½-inch copper pipe bent in an elongated S shape. Smooth joints don't collect water and, thereby, resist weathering and cracking.

Wear a pair of sturdy work gloves. The combination of wet mortar and rough concrete block can abrade the skin on your hands.

Concrete Building Blocks and Anchors

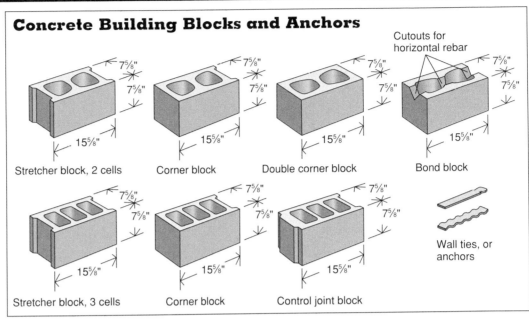

Stretcher block, 2 cells

Corner block

Double corner block

Bond block

Cutouts for horizontal rebar

Stretcher block, 3 cells

Corner block

Control joint block

Wall ties, or anchors

Preliminary Planning Sketch

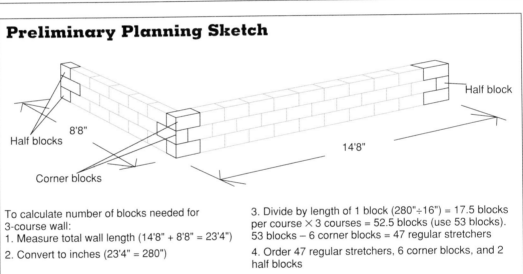

Half block

Half blocks

Corner blocks

8'8"

14'8"

To calculate number of blocks needed for 3-course wall:
1. Measure total wall length (14'8" + 8'8" = 23'4")
2. Convert to inches (23'4" = 280")

3. Divide by length of 1 block (280"÷16") = 17.5 blocks per course × 3 courses = 52.5 blocks (use 53 blocks). 53 blocks − 6 corner blocks = 47 regular stretchers

4. Order 47 regular stretchers, 6 corner blocks, and 2 half blocks

BUILDING A BLOCK WALL

Except for lifting the heavy blocks, building a block wall is relatively simple and the work goes quickly. If you want the wall to be decorative as well as functional, you can face it with brick, stone, tile, or stucco.

Preparing the Site

Good site preparation and a solid foundation will make the difference over the years in how your wall ages. First, a footing must be laid to support the wall, or it will soon crack and list as the earth moves beneath it from frost heaves or water saturation. Lay out the site using the techniques shown for a brick wall on page 34.

The footing should be twice as wide as the block and at least 6 inches thick. The technique for constructing a footing is described on page 58. Strike off the footing so that it is level to within ⅜ inch along its entire length. There is no need to float it, but it should be left to cure for several days.

Reinforcement is strongly recommended for any retaining wall and indeed may be required by your local building codes. The method described here is for reinforcing a retaining wall that is less than 3 feet high. Anything higher may have to meet stringent code requirements, including possibly having an engineer design the wall. Check with your local building inspection department.

The Dry Run

A dry run is essential in laying a block wall. It allows you to find and solve any problems before you start applying mortar, which makes mistakes difficult to correct.

As you place the blocks, use scrap pieces of ⅜-inch plywood to space the blocks ⅜ inch apart, the thickness of the mortar joint. Use mason's line to make sure that the blocks are aligned, then snap a chalk line on the footing exactly 2 inches away from the blocks. Use this chalk line as a guide; it will be far enough away not to be covered by mortar spread on the footing.

Laying the First Course

As you begin laying block, always place the side with the wider web edges up so that the block will hold mortar well. You will quickly get the hang of this because block is easier to pick up this way.

Note that the mortar is usually spread only on the edges of the block, not across the webs. Mortar is spread on

The last block laid in a course is called the closure block. Before sliding it into place, butter the inside of the corner block or both ends of the closure block.

How to Build a Concrete Block Wall

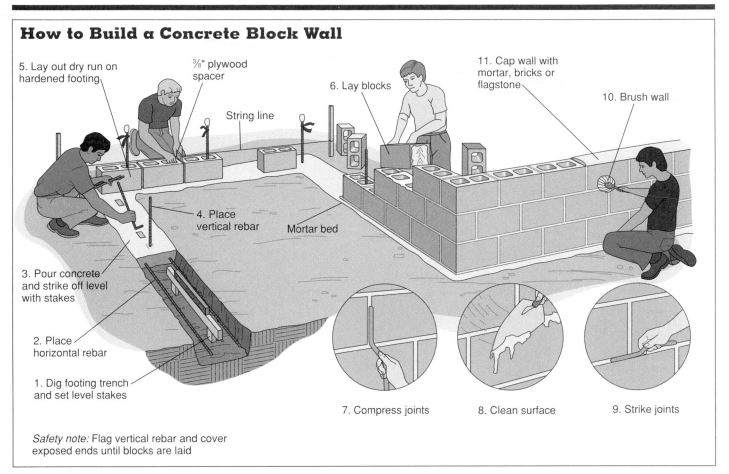

5. Lay out dry run on hardened footing

$\frac{3}{8}$" plywood spacer

String line

6. Lay blocks

11. Cap wall with mortar, bricks or flagstone

10. Brush wall

4. Place vertical rebar

Mortar bed

3. Pour concrete and strike off level with stakes

2. Place horizontal rebar

1. Dig footing trench and set level stakes

7. Compress joints

8. Clean surface

9. Strike joints

Safety note: Flag vertical rebar and cover exposed ends until blocks are laid

---

webs only if you want to markedly increase lateral strength, which is often done on low retaining walls without steel reinforcement.

Tip: If you plan to veneer the wall with stone or brick (see page 47), it is important to embed corrosion-resistant metal ties in the mortar about every 2 feet as you lay the courses of block.

Start laying the first course by spreading a solid 1-inch-thick layer of mortar on the footing for a distance of about 3 blocks. Spread the mortar about an inch wider than the block on each side. Position a block and press it down until the mortar is compressed to $\frac{3}{8}$ inch thick. Check the plumbness of the block with a carpenter's level.

Using a trowel with mortar on it, butter one end of another block and press it into place against the first block. Repeat with all the block in the first course. Level and plumb each block; also use the level against the sides of the block to check their alignment.

Adding Leads and Courses

Once the first course is completed, start building the leads (corners). Remember that every other course after the first course should start and end with a half block.

As you build each lead, use the level constantly to be sure that your work is level and plumb. In a well-

constructed lead, you should be able to place a straightedge between the top and bottom block and just touch each block in between (see illustration, page 36).

To lay the second course of block, spread a $\frac{3}{4}$-inch layer of mortar on top of the first course for a distance of 2 or 3 blocks. Position the blocks in this course as described above, buttering one end only. This can be done efficiently by standing 2 or 3 blocks on end and buttering them together with about $\frac{3}{4}$ inch of mortar.

Place the first block in this course. Then place the second block, keeping the buttered end raised slightly, then lowering it and fitting it snugly against the preceding block in one smooth motion.

Use the trowel handle to tap the block level, then use the blade to scrape away the excess mortar that is squeezed out. Mortar that drips on the block is better left alone to harden before you scrape it off; trying to remove soft mortar can result in smears that are difficult to clean.

Laying to the Line

With the leads completed, begin laying the block to the line. To keep the courses level and straight, stretch mason's line even with the top of the course to be laid. Each block should be even with the top of the line and about $\frac{1}{16}$ inch (about the thickness of the line) away from the line. Don't let the blocks touch the line.

As before, butter the tops of 2 or 3 blocks and the ends of the same number of blocks, and set them in place. Scrape off any mortar that has squeezed out and work it into the mortar on the board. If the position of any block needs adjusting, do it while the mortar is still wet; trying to move a block after the mortar has stiffened will break the bond.

Placing the last block, or closure block, is the trickiest part of laying block to the line. Expertise at this will come only with experience. Butter both ends of the closure block and both ends of the abutting blocks that are already in place. Hold the closure block directly above the opening so it is perfectly centered, then push it down with one smooth motion. Immediately scrape off the excess mortar, then tap the block until it is level and straight. If the mortar falls off the block on your first attempt, pull out the block, remortar everything, and try again.

Control Joints

In block walls 60 feet or longer, control joints should be established every 20 feet to minimize the chance of the wall cracking because of temperature changes. These control joints are vertical breaks in the wall that allow the wall sections to move up and down but still maintain lateral rigidity.

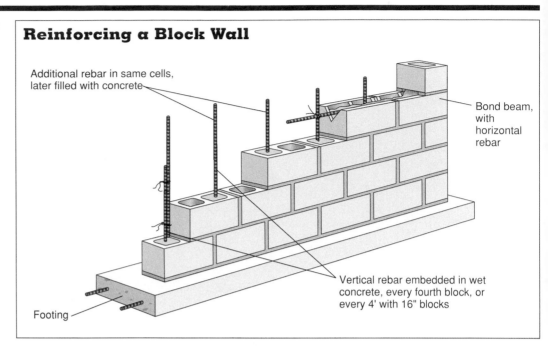

Reinforcing a Block Wall

Additional rebar in same cells, later filled with concrete

Bond beam, with horizontal rebar

Footing

Vertical rebar embedded in wet concrete, every fourth block, or every 4' with 16" blocks

There are three types of control joints you can use, depending on what is available in your area. One type, shaped like a rubber cross, fits into the grooved ends of special block. In another type, the ends of the block at the control joint are tongue and groove, one locking into the other. The third type, which does not require any special block, is known as the Michigan joint.

The Michigan control joint is formed by slipping a piece of roofing felt or building paper between 2 blocks where they meet. No mortar is used on the ends of these blocks, and the paper prevents any bonding as bed mortar is squeezed up. However, concrete is placed in the partial cell between the end of one block and the paper to provide lateral strength in the wall. After the wall is finished, the control joint is filled with concrete caulk, which is available at hardware or masonry supply stores.

Finishing Mortar Joints

As you work, keep scraping the excess mortar from the joints and reusing it by mixing it with mortar on the board. Once the mortar in the joints has hardened so that it will just barely accept a thumbprint with firm pressure, you must start tooling it.

First use a convex or V-shaped jointer to compress all the joints. This will force mortar out beyond the edge of the block; trim off this excess with the edge of the trowel.

Let the mortar in the joints set up a little more, then re-tool the joints—the vertical ones first, then the horizontal ones—to form a pleasing, distinct joint.

Finally, when the mortar has dried, brush the wall with a wire brush or other stiff brush to remove any dirt or small fragments of mortar.

Reinforcing a Block Wall

On long, straight retaining walls, reinforcing buttresses—called pilasters—can be built into the wall for additional strength. Since these blocks must also be supported by a concrete footing, plan ahead to widen the footing at these points when designing the wall and footing. Even greater strength can be achieved by filling the block pilasters either with concrete or with reinforcing rods and concrete.

Retaining walls are more usually reinforced with rebar in the wall blocks themselves. Lay out the blocks beside the footing trench in a dry run, spaced ⅜ inch apart to allow for the mortar joint.

Then drive ½-inch rebar into the trench every 4 feet in the center of where a block cell will fall. Make sure that the cells line up with the rebar. Remove the rebar from the trench and mark its location with wood pegs on the side of the footing ditch.

After the footing has been poured and begins to set up, push rebar into the concrete to about half the depth of the concrete (do not let it reach into the dirt). To get a better bite, try bending an L or J at the bottom of the bar before pushing it into the concrete.

Make sure that the rebar is plumb as the concrete continues to set up. The tops of the rebar need only be even with the top of the second course of block and can be cut to height after the footing has cured.

After the successive courses are laid, drop additional lengths of rebar down those cells that already have rebar. Fill each cell with concrete, then tamp it down with a stick to make sure that the cell is filled.

Capping a Wall

A garden or retaining wall of concrete block is not complete until it is capped. The cap not only gives the wall a finished appearance but also prevents moisture from entering it. The cap can be applied in any of several different ways. It may be applied before or after a veneer, depending on the type of veneer you have chosen.

A concrete block wall is usually capped with flat concrete block, which is simply mortared in place on top of the wall. In this case the mortared joints should be kept flush rather than tooled, which would cause a slight depression that would catch water.

For a more decorative finish, a block wall can be capped with stone, such as flagstone. The cells of the top course of

block are filled with pieces of scrap block and concrete before the stone cap is mortared in place. The stone should be set for a ¾-inch overhang so that water will drip off the cap rather than enter the wall.

If the wall is to be stuccoed, as described below, a common method of capping is to cover the wall with a rounded layer of concrete. The top course of block must first be filled with concrete. One way to do this is to cover the top of the second last course of block with wire mesh or roofing felt to prevent the concrete on the top from falling through to the bottom of the wall.

Be sure that the mesh or felt just covers the webs and not the edges of the block. While the mortar in the cells is still fresh, begin laying on the top layer of concrete, building it up in the center and then rounding it over the top of the wall with a trowel, much like smoothing frosting on a cake.

Applying Veneer

After the wall has cured for at least seven days, you can face it with brick, tile, stone, or stucco. Such a veneer can turn a plain wall into an attractive element in your landscape. Metal ties should have been embedded into the mortar of such walls during construction. The ties help keep the veneer attached to the wall (see illustration at right).

Applying Brick or Stone

For details on working with brick and stone, see pages 35 to 36 and 58 to 60, respec-

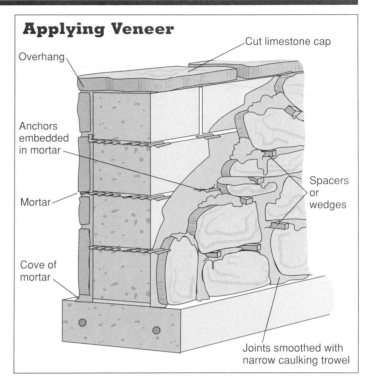

Applying Veneer

Overhang

Cut limestone cap

Anchors embedded in mortar

Mortar

Cove of mortar

Spacers or wedges

Joints smoothed with narrow caulking trowel

Block does not have to be boring, as this wall covered in vibrant rose-colored stucco attests.

tively. Once you've selected the veneer that will complement your landscape, applying it is easy.

Brickwork is the easiest; you are basically building a single-wythe wall (one brick thick). Use the same technique explained on pages 35 and 36. The only difference is that you must embed the wall ties in the brick mortar, which may require bending them to align properly.

To use flagstone, first do a dry run, using the biggest and heaviest stones for the first course. Then spread a generous bed of mortar onto the blocks and place the first course of stone. As you build up the wall, trim the stones to fit, then check their placement before applying mortar to the back of the stones and setting them in place. Try to set the stones so that water runs down the joints instead of puddling in them.

Once the stones are in place, apply mortar to the joints. Place wood wedges, or spacers, in the joints to keep the stones from squeezing out the mortar. Take care not to mortar the spacers in. After the mortar sets up, you can remove the spacers and plug the holes with additional mortar.

Clean the mortar from the stones as you go. When the mortar accepts a thumbprint with firm pressure, you can finish the joints.

Applying Tile

Tile can be applied to a rough concrete block wall that is clean, stable, and without any cracks. The tile you choose must be for exterior use and able to withstand humidity

Top: Block walls can be veneered with stone flags or lightweight, manufactured stone, embedded in a layer of mortar. Bottom: Block can be painted with exterior latex paint formulated for masonry. If you wish to smooth the surface before painting it, first apply a heavy-bodied block filler.

and temperature changes. Glazed tile is a common choice.

Do a dry run first against a snapped plumb chalk line. Determine where the tile will have to be cut (the fewer cuts the better). The cut tiles can be either at the top of the wall, under the cap, or at the bottom of the wall. Rent a power masonry saw for clean, sharp cuts, or use a masonry blade on a circular saw. If the wall is butting up against another wall or a building, place the cut tiles at the end of the run.

Apply a thin bond coat of mortar or thinset tile adhesive about ⅛ inch thick on the wall and let it cure. Then lay another coat about ¼ inch thick in which to set the tile.

After the tile has cured for at least two days, apply a weatherproof grout according to the manufacturer's directions, and seal it.

Applying Stucco

Stucco is similar to concrete but doesn't contain any gravel. Unlike stuccoing a house, stuccoing a wall does not require that wire be stretched over the block to hold the stucco. If your stuccoing efforts are somewhat rough, that only adds to the rustic charm of the wall.

Before applying the stucco, paint the wall with a concrete bonding agent, which can be found in most hardware stores (see illustration above).

Prepare the stucco mix from 1 part portland cement, 1 part hydrated lime, and 6 parts fine sand. Alternatively, you can use one part masonry cement, which already contains hydrated lime, and 3 parts sand. In either case, add

Applying Stucco to a Block Wall

1. Coat blocks with bonding agent

2. Apply scratch coat with plasterer's trowel

3. Score scratch coat with scarifier

4. After 2 days apply finish coat with plasterer's trowel

water until the mix has a rich, creamy texture.

Stucco is applied in two separate coats: the scratch coat and the finish coat. The latter can be mixed with a coloring agent.

Applying the Scratch Coat

Trowel on the first coat with a plasterer's trowel rather than a pointed trowel. Work down from the top of the wall, pressing the mortar firmly against the block and smoothing it out until it is just ⅜ inch thick.

Before the stucco dries, score it about ¼ inch deep so that the second coat will bond tightly to the first coat. Simply scratch the wall with several pieces of wire (cut-up pieces of a metal coat hanger will work) tied together and then spread apart, something like a broom.

Let the first coat cure for 48 hours either by repeatedly spraying it with water or by covering the wall with

plastic that is weighted at the ends and the ground to trap moist air.

Applying the Finish Coat

The second coat should be applied in the same fashion, making it ⅜ inch thick but leaving it smooth. Using a plasterer's trowel will make the work easier. This coat can be colored by adding a coloring agent to the stucco as you mix it, which eliminates the need for paint. The coloring agent should come with complete instructions for adding it to the stucco.

Building an Interlocking Block Retaining Wall

A retaining wall can be built of interlocking concrete block made specifically to hold back earth. There are many manufacturers of such block, and each has a different system of interlocking the block (see illustrations at right).

Retaining walls up to 3 feet high usually don't need building inspection approval; for higher walls, check your local codes.

Begin by staking out and stringing the area you want to retain. Then excavate a trench about 1 foot deep and about 16 inches wide plus the width of the block.

Add about 6 inches of well-compacted coarse gravel. Lay the base course of block below grade in the trench, or as recommended by the manufacturer. It is important that you level and align each block so that the rest of the wall will be straight. If the blocks have cells, fill them with gravel.

Place a layer of weed-blocking fabric on top of the gravel and against the soil you want retained.

Because water trapped behind a retaining wall often leads to later failure, make sure that you provide a way for water to escape. Perforated drainpipe can be installed on top of the weed-blocking fabric and behind the base

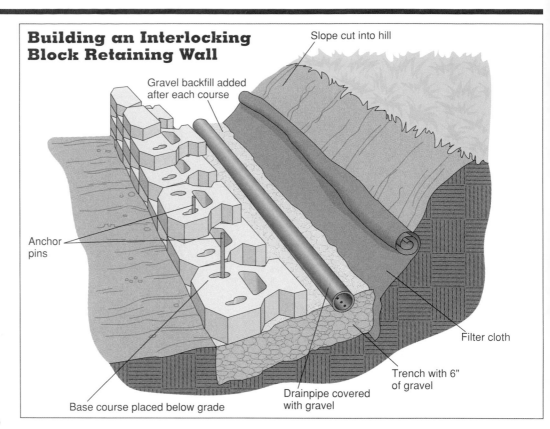

Building an Interlocking Block Retaining Wall

Slope cut into hill

Gravel backfill added after each course

Anchor pins

Filter cloth

Trench with 6" of gravel

Drainpipe covered with gravel

Base course placed below grade

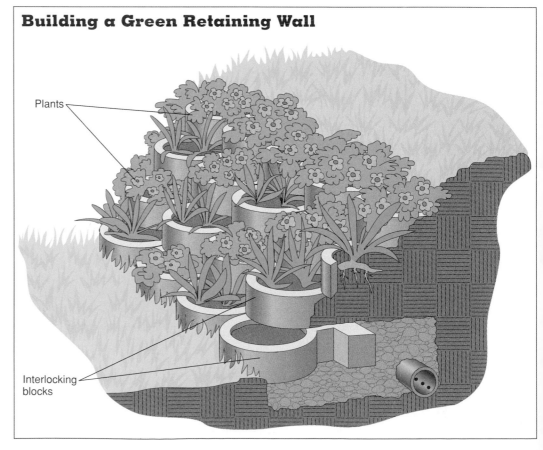

Building a Green Retaining Wall

Plants

Interlocking blocks

course of block; the pipe should be covered with plenty of gravel. See page 38 for more information on installing drainpipe.

Now you are ready to lay the first course of above-grade block, set back as per manufacturer's recommendations. Backfill with gravel. Set another two courses and backfill again with more gravel. On the last one or two courses, pull the weed-blocking fabric over the gravel, then backfill with dirt and tamp it down thoroughly.

At the beginning and end of retaining walls that are built into a slope, there should always be a return built into the hill. This return will retain the backfill material at the ends. Walls that follow a slope don't need a return.

Top and bottom: The blocks in this mortarless retaining-wall system are held together with reinforced fiberglass pins inserted through holes in the blocks.

USING STONE

The oldest building material on earth, stone blends well with almost any landscape. A flagstone walk or patio or a rough stone wall along a low bank is natural in appearance and complements virtually any style of house.

Working with stone is satisfying, from recognizing just the right stone for a particular spot to seeing the project completed. The more you work with stone, the more you will develop a flair for finding and fitting just the right piece in your project. A stone structure isn't just assembled, as brick or block projects are, with both materials and process repetitious. Because of the random nature of the material, working with stone requires continual selection, which means reassessing your supply each time you choose another stone.

In this chapter you will learn the basics of finding and working with stone, and how to build a stone wall, patio, and walk.

Because building with stone involves such basic elements as a few sacks of mortar, some simple tools, and the stones, the work itself can be as enjoyable and rewarding as the results. Just allow plenty of time.

STONE BASICS

Working with stone requires fewer tools than most other masonry projects. When you work with stone, you can proceed at your own pace and quit when you have had enough for one day. The stones have been here for thousands of years; they can wait until tomorrow.

Kinds of Stone

All rock is formed in one of three ways. Sedimentary rock is formed from deposits of sand, clay, gravel, and even plants and animals laid down by the action of water. These sediments pile up in horizontal layers and, over millions of years, solidify under the pressure and chemical changes. Sedimentary rock covers more than two-thirds of the earth's surface. The most common types are limestone, shale, and sandstone.

Metamorphic rock derives from existing rock that has been exposed to temperatures and pressures higher than those of its initial environments of formation. This exposure causes a chemical transformation in the rock that brings out new crystals and minerals, similar to what happens to clay when it is heated in a kiln.

Limestone turned into marble, shale became slate, soft (bituminous) coal became hard (anthracite) coal, and so on.

Igneous rock is usually formed by volcanic action. Some types are soft, such as lava, but many types are hard, such as granite and basalt. These types of rock can be cut and polished, and have excellent building qualities.

Stone used by stonemasons comes from quarries (quarried stone) or is found on the ground (fieldstone). Quarried stone requires considerable labor, which makes it costlier than fieldstone. Quarried stone includes marble, flagstone, and slate. Stone that is cut and shaped at the quarry site but has its face left natural is called ashlar.

Fieldstone, as the name implies, is found in fields and woods and along rivers and streams. Fieldstone is divided into two classes: rubble and roughly squared. Rubble is unworked stone in its natural state. Roughly squared stone has been worked over with a hammer and stone chisel to roughly shape it, which facilitates fitting one stone against another.

Where to Find Stone

You can buy stone from stoneyards or suppliers of landscaping materials. Compare the stones, prices, and delivery costs of several different sources. Cut stone, such as ashlar and flagstone, is the most expensive. Roughly squared stone costs less but will still be more expensive than ordinary rubble. If you are buying rubble, check that the stones are relatively flat rather than round.

If you want to collect your own stone, the easiest place to look is along rivers and streams. You may be able to salvage stone from an old foundation or the collapsed chimney of a long-gone house. Before removing any stone, be sure to obtain permission.

Caution! Fieldstone often shelters bees or snakes. Always lift the stone with a long pry bar to let whatever may be lurking escape.

Moving Stone

Collecting your own stone is hard work, so bring plenty of helpers. You'll need a long pry bar, a crowbar, and an understanding of leverage. Remem-

The skillful use of stone in this landscape—the fieldstone boulders, cut stone pavers, and crushed rock—introduces both formal and informal elements into the design.

Moving Large Stones

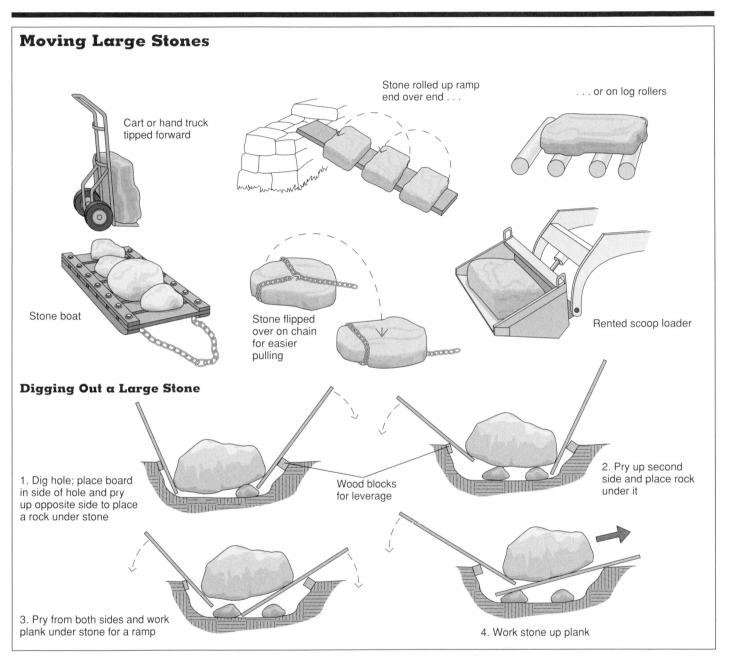

Cart or hand truck tipped forward

Stone rolled up ramp end over end . . .

. . . or on log rollers

Stone boat

Stone flipped over on chain for easier pulling

Rented scoop loader

Digging Out a Large Stone

1. Dig hole; place board in side of hole and pry up opposite side to place a rock under stone

Wood blocks for leverage

2. Pry up second side and place rock under it

3. Pry from both sides and work plank under stone for a ramp

4. Work stone up plank

ber, always lift with your legs, not with your back (see illustration on page 14). Keep your work low to the ground.

If you plan to salvage a lot of fieldstone and foundation rubble, buy or build a stone boat that you can chain to a trailer hitch and drag across a field. A stone boat resembles a sturdy sled with wide iron or steel runners. Since the stone boat is low to the ground, all you have to do is push or roll the stones onto the boat (see illustration above).

Once you've assembled the stones in one place, you can transport them to the building site on a low trailer. Use a long, sturdy ramp up which you can roll the stones, either on log rollers or by flipping them end over end. To move stones too big for rollers, rent or borrow a flatbed truck with a hoist, or rent a small tractor with a scoop loader.

You can move stones that are small but still too heavy to carry by using a gardening cart or a child's metal wagon tipped on end so that you can push or roll the stones into it. A hand truck will also work, but its small wheels may get bogged down in soft dirt, so bring some boards on which you can roll it. If you are load-ing stone into a wheelbarrow, put it toward the back to keep the weight off the front wheel.

Estimating Stone

The amount of stone you need depends on the project and the type of stone you plan to use.

If you are buying stone, first calculate either the square footage for a patio or walk, or the cubic feet for a wall. To calculate square feet, simply multiply the length times the width. With that dimension, the stone yard can tell you how much material you will need. When ordering cut stone, add 10 percent to your total to allow for breakage.

To calculate cubic feet for building a wall, multiply the length times the width times the height. Stone yards often sell stone by the cubic yard. To find how many cubic yards are in your wall, divide the cubic footage by 27.

If you are building a rubble wall from collected fieldstone, it is difficult to know when you have accumulated enough stone. Try to have more on hand than you think you will need, because much of the stone you have simply won't fit together. If you are buying rubble, add 25 percent to your order to allow for stone that won't fit.

Working With Stone

Stone is shaped by chipping, cutting, or splitting. It takes a while to get the hang of these skills, so practice first on waste rock.

Tools

Depending on the type of stonemasonry you plan to do, you will need some or all of the following tools.

• Heavy stonemason's hammer (5 to 8 pounds). Provides the necessary striking weight. An ordinary hammer is too light and may chip easily.

• Striking hammer. For driving wedges and chisels.

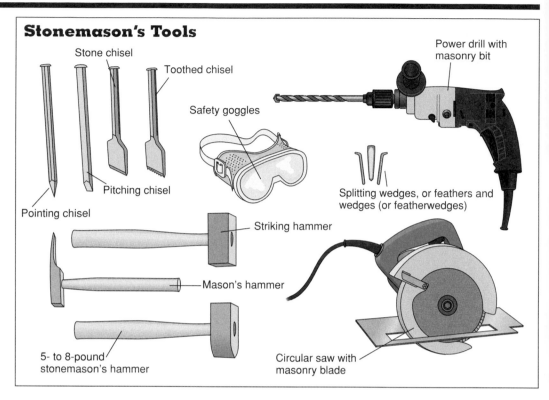

Stonemason's Tools

Stone chisel
Toothed chisel
Safety goggles
Power drill with masonry bit
Pitching chisel
Pointing chisel
Splitting wedges, or feathers and wedges (or featherwedges)
Striking hammer
Mason's hammer
5- to 8-pound stonemason's hammer
Circular saw with masonry blade

• Mason's hammer. Used to chip off corners.

• Stonemason's chisels. Include a pitching chisel, which resembles a brickset but is considerably heavier, a pointing chisel, and a toothed chisel.

• Protective eyewear. Essential when cutting stone. Choose shatterproof safety goggles that fit firmly at the sides of your eyes or safety glasses with corner shields.

• Gloves. Wear heavy leather work gloves when cutting and lifting stone.

• Shovel. Square-nosed type, for scooping sand, gravel, or concrete.

• Wheelbarrow. Heavy-duty, or contractor's, type.

• Cement trowel. Large brick trowel, for working with mortar.

• Pointing trowel. Small triangular trowel, for filling joints with mortar.

• Power drill and masonry bits. For drilling holes.

• Circular saw with a masonry blade. For scoring fieldstone (wear a dustmask).

• Feathers and wedges (also called featherwedges, or splitting wedges). For splitting stone. Two short wedge-shaped metal rods (feathers) are inserted into a hole and forced apart by another rod (the wedge) driven between them. Featherwedges are difficult to come by but worth the search.

Cutting and Shaping Stone

Stone doesn't always need to be cut, particularly if you are building a dry stone wall, but cutting will help considerably in fitting stones together (see illustration opposite).

Cutting Fieldstone

When cutting fieldstone, place the stone to be cut on a surface that is firm but resilient, such as the ground, a bed of sand, or scrap pieces of wood. Don't place the stone on concrete or other stones; it may crack in the wrong place.

Once you have decided where to cut, usually at some protrusion or sharp irregularity that you want to remove, mark a line with chalk, crayon, or a soft pencil. If a stone has a natural fissure, make your line there, because the stone may break there anyway. Then score the line with moderate blows of a hammer and chisel. Turn the stone over and repeat the process on the other side. Now place the chisel on the mark and strike it sharply with the hammer. Not all stone will crack; set aside the ones that won't and try others.

Cutting Stone

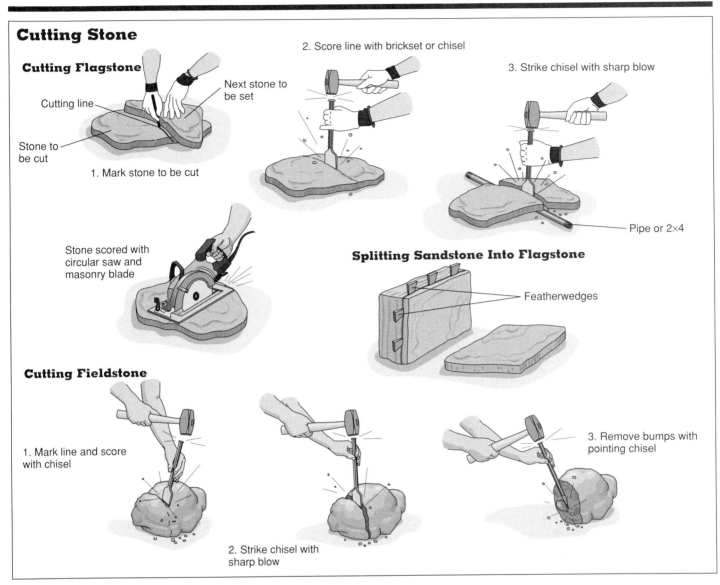

Cutting Flagstone

Cutting line

Next stone to be set

Stone to be cut

1. Mark stone to be cut

2. Score line with brickset or chisel

3. Strike chisel with sharp blow

Pipe or 2×4

Stone scored with circular saw and masonry blade

Splitting Sandstone Into Flagstone

Featherwedges

Cutting Fieldstone

1. Mark line and score with chisel

2. Strike chisel with sharp blow

3. Remove bumps with pointing chisel

After the cut is complete, dress the edge (remove bumps and sharp irregularities) with a pointing chisel. Place the point of the chisel at the base of the bump, then give it sharp raps with the hammer.

Cutting Flagstone

First, mark a cutting line and score it with a chisel. Then place the stone over a length of pipe or scrap 2×4, and strike sharply on the scored line with a hammer and chisel.

To cut a curve that matches the edge of an adjoining flag-stone, place the stone to be set over the stone to be cut and trace the outline of the curve with a pencil. Score the line and cut as described above.

Another method for scoring flagstone is to use a circular saw with a masonry blade. Wear a dust mask. Carefully position and brace the stone. Firmly hold the saw with two hands and lightly make repeated passes to deepen the score. To complete the cut, hit the stone at the scored line with a stonemason's hammer.

Splitting Large Stones

You can split a large flat stone, 3 or 4 feet across, with several pairs of feathers and wedges. Try a local rental center, a masonry supply yard, or concrete tool supplier (feather-wedges are also used to split concrete).

First, draw a chalk cutting line. Then, using a power drill with a masonry bit the diameter of the featherwedge assembly, drill a hole every 8 inches along the line.

Insert the featherwedges into each hole. Starting in the middle, then working alternating sides, gently tap each featherwedge. The pressure should cause the rock to split along the line. If the rock doesn't split, redrill holes at 4-inch intervals and try again. Be careful not to jam the wedges into the feathers so tightly that you can't pull them out. If that happens try using a crowbar to pry them out. Place a wood block under the crowbar so it does not mar the stone.

WEEKEND PROJECTS

The most common uses of stone for backyard projects are for walls, patios, and walks. Although each type of project involves different stoneworking techniques, once you acquire a few basic tools and skills, you can adapt them to any project.

Building a Stone Wall

You can build a stone wall with or without mortar, depending on the desired appearance and strength.

A stone wall should be approximately 2 feet thick for every 3 feet of height. For every 6 inches you raise the wall beyond that, you should thicken the base 4 inches. The height of your wall may be regulated by local building codes. Walls 3 feet or less in height usually don't require a permit. Higher walls may require a permit and possibly even an engineering study, so check with your building inspector. If you are building a wall on a property line, be sure to check the boundaries.

Building a Mortared Stone Wall

A mortared stone wall is a showpiece in any yard or garden. If you have the patience and time, building such a wall is a rewarding endeavor (see illustration opposite).

Preparing the Site

A mortared stone wall is inflexible, so it should be supported by a solid footing. Otherwise, it may crack due to settling or frost heaves.

To prepare the footing trench, first lay out the wall site with stakes and string. The footing should be 6 inches wider than the base of the wall.

The footing should be at least 6 inches thick. In areas where the ground doesn't freeze in winter, the footing trench should be 8 inches deep, so that the top of the footing will be 2 inches below the ground level. In cold climates, the footing trench should be dug below the frost level. In addition, footings should have two lengths of ½-inch rebar placed in the middle to prevent the concrete from cracking. For more instructions on working with concrete, see pages 69 to 76.

If the stone wall will be backed by a moderate amount of earth, as in a planting box, drainage holes must be incorporated into the wall so that water can escape from behind the wall. The holes can be as simple as unmortared gaps left along the base.

If the wall will be backed by a considerable amount of earth, as in a retaining wall, a more complicated drainage system must be installed, using gravel, drainpipe, and weep holes. For a detailed description of how to incorporate draining into a retaining wall, see pages 38, 50, and 51.

Fitting the Stones

The key to building a sturdy, pleasing wall is to carefully select and fit each stone. The only way to do this properly is to dry-fit each stone before applying mortar. Avoid using an awkward-sized stone with the idea that the mortar alone will hold it in place. Gravity will eventually prevail. Small stones can be wedged between large stones to help them fit.

The low, stacked stone walls around this patio help to define the edges and provide a transition to the natural landscaping.

Building a Mortared Stone Wall

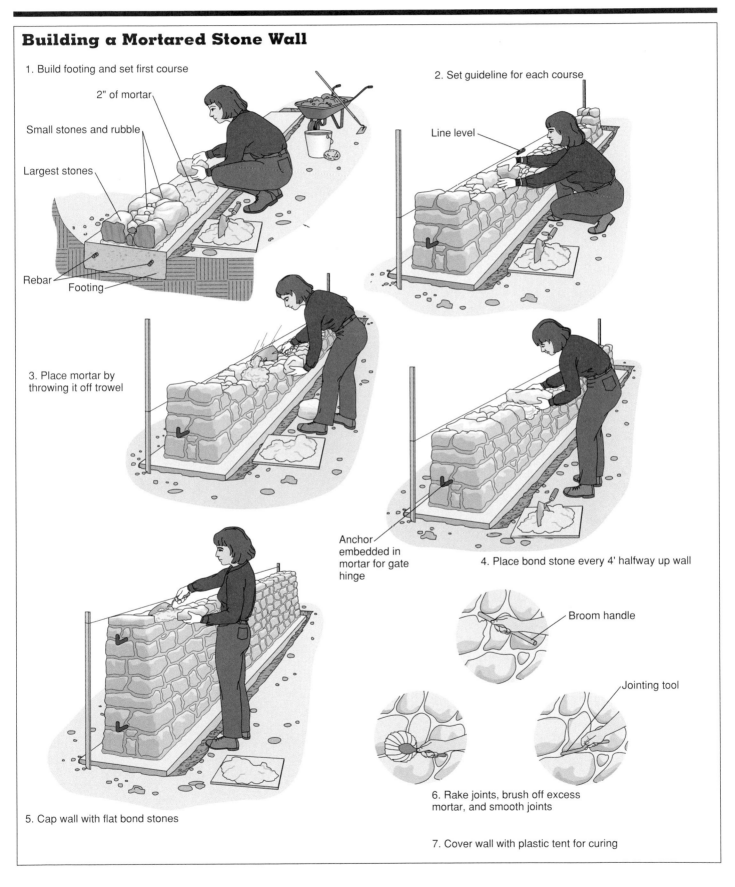

1. Build footing and set first course

2" of mortar

Small stones and rubble

Largest stones

Rebar

Footing

2. Set guideline for each course

Line level

3. Place mortar by throwing it off trowel

Anchor embedded in mortar for gate hinge

4. Place bond stone every 4' halfway up wall

Broom handle

Jointing tool

5. Cap wall with flat bond stones

6. Rake joints, brush off excess mortar, and smooth joints

7. Cover wall with plastic tent for curing

Laying the First Course

Start the first course by arranging the largest stones on the footing, turning them back and forth to get a good fit. Keep in mind that the wall will be at least 2 stones thick; the center portion will be filled with small rubble and mortar.

Now remove the first course of stones, spread a 2-inch-thick layer of mortar on the footing, and set the stones back in place.

Building the Ends

After you have set the first course on the footing, start building up the ends. Stretch mason's line between 2 stakes driven at each end of the wall; set the line about 3 to 4 inches above the top of the next course. Use this line as a rough guide to keep the wall level and plumb.

Use the flattest stones with the smoothest faces for the wall ends, and interlock the stones as much as possible. Save some of the flattest stones for capping the wall.

Laying the Middle Courses

The motto for a good stonemason is, "One rock over two, two rocks over one." Remembering this will help you avoid aligning joints directly over one another, which weakens the wall.

Begin the second course of stone by dry-fitting several stones at a time until you have a smooth and stable fit. Then remove the stones, apply the mortar, and fit the stones back into place. When fitting two stones against each other, spread mortar on the one already in place.

As you work your way higher on the wall, periodically lay bond stones across the wall to tie it together. Place these about 4 feet apart horizontally and approximately halfway up the wall, or more often if you have good bond stones. Raise the wall by only one or two courses per day; the weight of additional courses will force mortar out of the lower joints.

Mixing the Mortar

The mortar used to lay stone is similar to that used to lay concrete block. Stone is heavier, however, so the mortar must be made a bit stiffer by using a little less water. Also, you should add hydrated lime, which makes the mortar bind better and reduces the mortar stains on the stone.

A standard mortar mix for laying stone is 1 part portland cement, 1 part hydrated lime, and 6 parts sand. Mix the mortar in a wheelbarrow as described for laying brick (see page 23). Add enough water to moisten the mortar mix: It should not be creamy, as is brick mortar. When you trowel on stone mortar, it should stay in place, not ooze down the side of the stone.

Applying the Mortar

It is important before you apply mortar to clean off the stone of any dirt, moss, or sand, which would prevent the mortar from bonding well to the stone. Either brush the stone or hose it off and let it dry. Applying mortar to wet stone risks oozing and staining.

Place a couple of shovelfuls of mortar on a piece of ply-

wood and keep it on the ground beside you. Don't be too delicate when placing the mortar; instead, throw it from the trowel onto the stone.

Work the mortar with the tip of the trowel to settle it into the opening, then set the stone in place. Rap the stone firmly with the end of the trowel handle to settle it and force out any air bubbles. Scrape off the excess mortar and throw the scrapings into the center portion of the wall.

Mortar will begin to set in about 30 minutes, so spread only as much as you can use in that amount of time. If the mortar in the wheelbarrow gets too stiff while you are getting started, add a little water. To keep the stones clean as you work, have a bucket of water with a large sponge handy to wipe off spills immediately.

Tip: If you don't want new-looking mortar joints, tint the mortar to gray or another dark earth tone with a powdered tinting agent that is added to the mortar mix; it is available at masonry supply stores.

Jointing the Wall

The mortar in the face joints of the wall can be left flush with the stone or raked out. The indentations of raked joints create shadows that add definition to the wall.

You can remove up to ½ inch of mortar in the joints without weakening the wall. This is most easily done with the end of an old broom handle or a piece of ¾-inch pipe.

The mortar is ready to be raked when only a shallow thumbprint remains when the mortar is pressed firmly. This is about half an hour after the

mortar has been applied. If you are building a long wall, be careful to joint the first part before the mortar is too hard.

Rake the joints to the depth you prefer. Then go over them with a soft brush, such as an old paintbrush, to remove the excess mortar. For a smoother joint, go over them again with a jointing tool.

Capping the Wall

An excellent way to finish a wall is to mortar a row of flat bond stones across the top of the wall from front to back, or use contrasting cut stone, such as granite or slate. Or you can simply save the flattest stones you have and mortar them in place along the top of the wall. For a finished look, leave the flat stones flush with the top edges of the stones beneath them instead of raking these joints.

Building a Dry Stone Wall

It is especially important with a dry stone wall to taper it inward at the top. This is called a battered wall.

Make a batter gauge from 1×2 lumber, or from any scrap wood, to guide your work. The rule of thumb for battering a wall is to tilt it back 1 inch for every 2 feet of rise.

A stone wall should be about 2 feet thick for every 3 feet of height. If you are building a low retaining wall (2 to 3 feet high) against an existing bank, however, the wall need be only about a foot thick. If your stones are about a foot in diameter, the wall can be just one stone thick. In such a case, it is a good idea to first batter the dirt bank (taper it into the slope) with a pick and shovel before building the wall. See below about whether a gravel base is needed.

Planning the Wall

When building a dry stone wall, you must carefully select each stone for the best fit. If a stone is not well anchored in place by surrounding stones, it will eventually fall out, and part of the wall will collapse.

As with a mortared stone wall, a dry stone wall is built by laying two parallel lines of stone and filling in the center portion with small rubble. As the wall goes up, long and flat bond stones must be placed periodically across the wall to tie the two rows together (see illustration at right).

Whether or not you need a footing to support a dry stone wall depends on your winter climate. If the ground freezes, you should dig a footing trench below the frost line and place the largest stones there as a footing. Or you can pour a concrete footing, as described for a mortared stone wall (see page 58), or lay gravel or crushed stone in the footing trench. Without a footing, frost heaves will likely topple part of the wall every winter. In mild climates a footing is optional.

Laying the Base

For a stone footing use your largest and flattest stones. Mark the perimeter of the wall, then excavate the ground about 4 inches deep, or below the frost line in cold areas.

Arrange the stones in two parallel rows in the excavation and remove or add dirt to stabilize them. The stones should

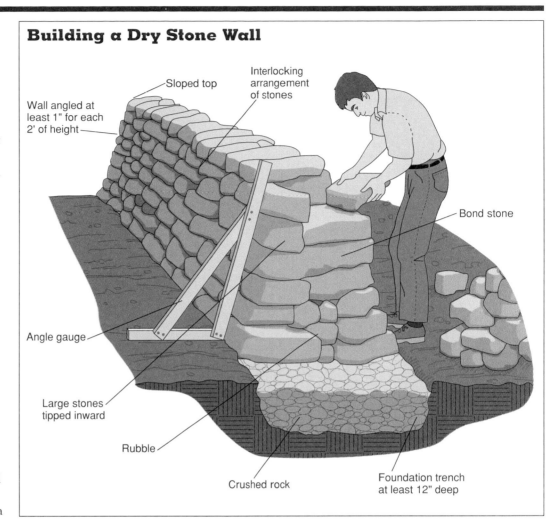

Building a Dry Stone Wall

- Sloped top
- Interlocking arrangement of stones
- Wall angled at least 1" for each 2' of height
- Bond stone
- Angle gauge
- Large stones tipped inward
- Rubble
- Crushed rock
- Foundation trench at least 12" deep

not wobble. They should not tilt out, either; this would cause successive courses to tilt away from each other, resulting in a weak wall.

Building the Ends

As with a mortared stone wall, the ends of a dry stone wall are built before the center section. Use your longest and flattest stones for the ends and corners. Place the stones so that each one interlocks with the others. Save some of your flattest stones for capping the wall (see right). The smoothest surfaces of the stones should face out to give the wall a finished appearance.

If you can't find stones that fit well enough, you can cut them. Take your time with this step, because good construction at this point is important to the stability of the entire wall.

Laying the Middle Courses

The remaining courses of stone should be laid as described for a mortared stone wall (page 60), using a guideline to keep the work level. Also check the face of the wall periodically with the batter guide and level.

When the wall is about half complete, remember to tie it together with bond stones

about every 4 feet horizontally and approximately halfway up the wall, or more often if you have good bond stones.

Capping the Wall

Laying a series of flat bond stones across the top of the wall not only gives it a finished appearance but also holds the wall firmly together. If you have a lot of irregularly shaped stones left over, fit them along the top of the wall. You can greatly increase the strength of the wall if you mortar this last course in place, as discussed for a mortared wall (see page 60).

BUILDING A FLAGSTONE PATIO

Flagstone is an ideal material for a patio. It's attractive, durable, and surprisingly easy to use. A sedimentary rock, it can be found naturally in many parts of the country, but more likely you will end up buying it—to have enough to complete the job.

A flagstone patio can be built in one of three ways: in a lawn, on a sand bed, or on a concrete base. First, spread out all the stones, just like spreading out all the pieces of a jigsaw puzzle. This gives you a head start to fitting the odd-shaped stones together.

In a Lawn

The simplest (and least durable) method of building a patio is to place each flagstone individually in a lawn. The grass makes a nice filler between the stones.

Arrange the stones on the grass, flattest side up. Then cut an outline of one of the stones in the grass with a trowel. Remove the sod and enough dirt so that the top of the stone is about 1 inch above the surface of the ground. If the stone wobbles, replace some of the dirt and pack it down until the stone is seated firmly. Repeat with all the stones.

Besides an uneven surface, there are two drawbacks to this type of patio. One is that you will have to cut the grass that grows between the flagstones. The other drawback is that the stones tend to settle, so after about five years you may have to reset them. This involves picking up the stones and placing sand under them to raise them to their original level.

In Sand

Building a flagstone patio in sand is very much like building a brick or paver patio in sand (see page 27). As with any paving on a sand base, it requires a solid edging of landscape timbers, bricks set in soldier courses, a concrete curb, or commercial-grade plastic edging.

The flagstones in this patio are actually large slabs of broken concrete. Instead of mortar between them, herbs have been planted to enhance the natural effect.

After installing the edging and building the sand base, set the perimeter stones tightly against the edging; try to use stones that have one flat side. Then, working in a circle, fill in the rest of the patio. Finally, fill the joints with sand, wet it, and tamp it.

On a Concrete Base

The most permanent flagstone patios are laid in a bed of mortar on a concrete slab (see illustration, page 65). This is a good way to upgrade a concrete patio or walk. If you need to build a new concrete base, see pages 80 to 83.

Before mixing the mortar, lay out the flagstone and decide on the placement. Take the time to arrange the stones to minimize the number to be cut. This also lets you see whether you have enough stone on hand. Then mark the position of the stones, remove them, and brush them clean.

If the flagstone is gauged (uniform thickness), install temporary screed boards around the perimeter of the slab to screed the mortar to a uniform thickness. The tops of the boards should be even, about 1 inch above the slab. If the stones are uneven, this step is unnecessary.

Laying the Flagstone

Mix the mortar, using 1 part portland cement to 3 parts masonry sand. Keep the mix fairly stiff so that it will support the flagstone. When you're ready to place the stones in the mortar, spread mortar on the slab in an area large enough to seat 2 or 3 stones. Spread the mortar about 1 inch thick, or more if necessary to firmly seat the stones.

Flagstones can be installed, top, over a concrete base and finished with mortar joints, or, bottom, over a base of compacted gravel and sand, with sand swept into the joints.

Space the stones approximately ½ inch from one another and from the edging. Set a stone, then rap it with the handle of the trowel to make sure that it is well seated. Because of the irregular surface of stone, it is impossible to set it perfectly level, but you can come close by laying a straight 2×4 across several stones at a time with a level on top of it to check your work. If a stone is too high, press it farther into the mortar. If a stone is too low, take it out and add mortar, then fit it back in place.

As you work your way toward the center, put down a piece of ½-inch plywood large enough for you to kneel on; it will distribute your weight evenly on the flagstones.

Let the mortar set for at least 24 hours before filling the joints. Firmly pack the mortar in the joints, then smooth it with a slicker tool or trowel. Carefully fill the joints, troweling in the wet mortar so that as little as possible falls on the flagstone. What mortar does fall on a stone should be wiped up immediately with a large, wet sponge rinsed frequently in a nearby bucket of water.

Curing the Mortar

For the mortar to cure properly, it should not dry out too rapidly. The best curing method is to cover the entire patio with sheets of plastic weighted at the edges and at all the joints. This traps the moisture and allows the mortar to cure slowly, which greatly increases its strength. Let it cure for about four days before walking on it.

Building a Walk

A simple stone walk through a garden or across a lawn is a pleasing design feature that is easy to build. Select stones that are large and flat.

In a Lawn

For a walk across a lawn, the first step is to lay out the walk with lengths of rope or garden hose (see illustration opposite).

Next, lay out the stones in a pattern you like. The walk can be single stepping-stones or clusters of stones so that two people can pass. Single stepping-stones should be placed a comfortable stride apart, not so close that you have to tiptoe or so far apart that you have to stretch for each one.

Set the stones in place, flat side up. Then, using a trowel, cut the sod in the exact outline of each stone. Remove the stone and sod, then dig a hole and reseat the stone, adjusting it until the top of the stone is about 1 inch above the lawn

The flagstones in this meandering walk are repeated as accents on the walls of the home, unifying the home with the landscape.

surface. It is usually necessary to put dirt back in the hole and pack it down until the stone is seated firmly and doesn't wobble. Repeat with all the stones.

If you are using clusters of stones for a wide walkway, space them about 4 to 5 inches apart so that a strip of sod remains between them, both for appearance and improved stability.

On Contrasting Gravel

Building a walk using large flagstones on a bed of crushed rock or rounded, light-colored pebbles is a favored technique of many landscape architects. First, lay out the walk, extending it about 1 foot beyond each side of the flagstone path

itself. (The overall width of the path depends on the scale you want to achieve, along with the size of the crushed rock or pebbles you are using.)

Select an edging material. Excavate down to the depth of it. Apply a preemergent weed killer, then lay down weed-blocking fabric.

Now install the edging and backfill it. Place your choice of crushed stone or pebbles between the edges of the path and to a depth that, when screeded, will be about ½ inch or so below the edging.

Then place the flagstones so that they are flush with the top of the edging. You may have to add or remove crushed stone or pebbles beneath the flagstones to ensure that they are seated well and don't wobble.

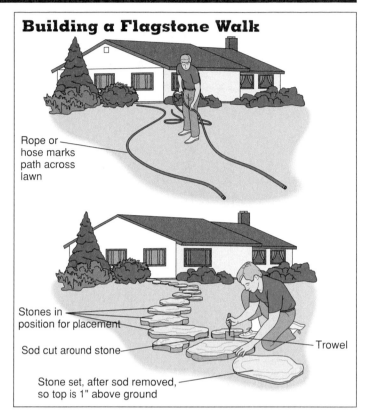

Building a Flagstone Walk

Rope or hose marks path across lawn

Stones in position for placement

Sod cut around stone

Trowel

Stone set, after sod removed, so top is 1" above ground

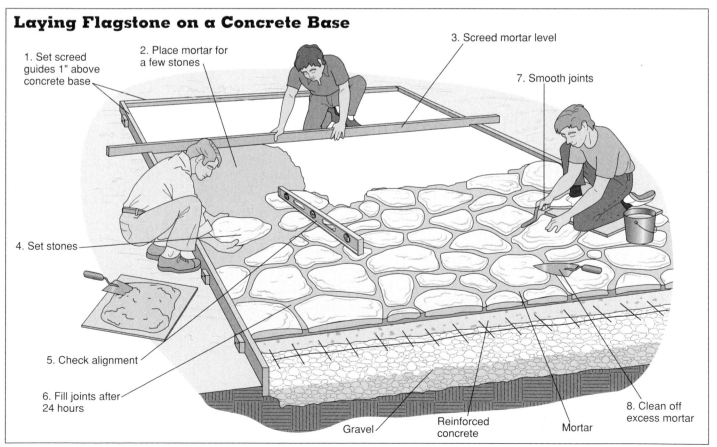

Laying Flagstone on a Concrete Base

1. Set screed guides 1" above concrete base
2. Place mortar for a few stones
3. Screed mortar level
7. Smooth joints
4. Set stones
5. Check alignment
6. Fill joints after 24 hours
8. Clean off excess mortar
Gravel
Reinforced concrete
Mortar

USING CONCRETE

Concrete is a versatile construction material for landscape work. It has numerous uses, from lowly footings for brick walls or deck posts, to decorative patio surfaces. It's inexpensive and relatively easy to work with, and can be shaped, colored, and finished in many ways. The biggest drawback to this medium is that mistakes are not easy to correct.

That's why this chapter includes techniques for all phases of concrete work, from layout through finishing and curing. All of the projects are within the capabilities of any do-it-yourselfer, even if you have never worked with concrete before. You will find complete information for building concrete patios, walks, slabs, stairs, low walls, footings, and foundations. Some of the projects can be completed in a few hours; others may require several weekends. Creating a concrete project not only adds beauty and value to your home, it gives you the satisfaction of building something durable and very useful.

If you are lucky enough to be pouring concrete where the ready-mix truck can back up to the forms, all you need to do is to have the driver move the chute around to place concrete where you want it. Most chutes extend up to 15 feet from the truck, but check ahead of time. Have a wheelbarrow handy to hold some extra concrete, should you need to fill in a low spot or two.

THE HARD FACTS

Concrete is a versatile building material, as suitable for colossal buildings as backyard stepping-stones. Understanding the fascinating characteristics of this remarkable building material will help you work with it.

What Is Concrete?

Concrete is a mixture of sand, gravel, cement, and water that hardens into a unified, stone-like mass. The glue that holds it all together is portland cement, a formula patented by an English mason in 1824. It was called portland because it had the same color as limestone quarried on the Isle of Portland, off the British coast. It is a measured mixture of limestone, clay, and gypsum that is pulverized, burned, and then ground. When it is mixed with water, a chemical reaction occurs, called hydration, that causes it to harden.

Concrete has tremendous compressive strength—the ability to withstand crushing—but it has very little tensile strength—the ability to withstand stretching. As a result, it has limited use unless it is combined with another material that has extraordinary tensile strength: steel. By embedding steel bars or mesh inside concrete, engineers and builders have created reinforced concrete, a material that is strong, durable, easy to work with, and inexpensive.

Engineers and chemists have developed many other ways to modify concrete for specific purposes. Besides altering the ratio of ingredients to give concrete different strengths, they have developed additives to accelerate or retard setup time, to prevent deterioration of the surface due to freezing and thawing, and to make it possible to work with concrete under such diverse conditions as extreme heat, extreme cold, or underwater. Controlling the basic ingredients, however, is still the key to successful concrete work.

Making Concrete

Making concrete is quite simple. The key is knowing how much of each ingredient to use.

Sand and gravel, known as aggregates, make up about 70 percent of the mix. A good concrete mix has enough small, or fine, aggregates to fill the spaces around the larger aggregates. Together, the various sizes of aggregate "nest" into a dense matrix that is held together by the cement.

Sand, which is either natural or made by crushing rock, is considered a fine aggregate. The sand should be clean, or washed, and the grains uniform. Never use beach sand; the salts will inhibit bonding and the concrete will weaken and deteriorate.

Gravel, or sometimes crushed stone, is a coarse aggregate. It ranges in size from ¼ inch to 1½ inches in diameter. An aggregate mix is sized by the largest aggregates in the mix. Most projects call for medium-sized gravel, which is ¾ inch. A general rule of thumb is that the aggregate size should not be more than one-third the thickness of the concrete slab.

Coarse aggregates should not have any chert, which is a type of rock that holds moisture. The use of chert can cause popouts in the concrete during freezing and thawing.

Aggregates added to concrete should be free of dirt, which prevents the cement from bonding to the gravel and also weakens the mix. Aggregates should be stored on sheets of plastic or plywood; this not only keeps them clean but also prevents waste.

Water, the catalyst that causes the cement to harden and bind to the aggregates, should be clean enough to drink. The amount of water needed will vary slightly with the purpose of the concrete.

The amount of water usually recommended is 6 gallons

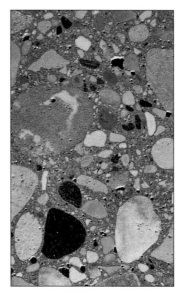

This cross section of concrete shows how the smaller sizes of aggregate and sand, or "fines," should fill the spaces between the larger stones to produce a dense mass.

per 94-pound sack of cement. (A sack of cement is not the same as a sack of ready-mix concrete, which contains the aggregates and requires much less water per sack.) Add water to the mix a little at a time. If you add too much, tiny subsurface air pockets can form, resulting in spalling (breaking off in chips or scales) and reduced strength.

Additives

Air-entrainment is an additive used in areas with severe cycles of wetting and drying or freezing and thawing. This type of concrete contains billions of microscopic air cells per cubic foot, which help relieve internal pressures and prevent cracking. Magnesium or aluminum tools should be used to bull float and finish this type of concrete, because wood floats tend to tear the surface.

Accelerators, water reducers, or retarders are often added to concrete, but this is better left to the experts. Before mixing concrete, ask your supplier's advice on what should be added.

Tip: The setting-up time of concrete may be retarded by adding sugar to it. If the motor fails on your mixer, add a pound of sugar; you'll have time to clean out and rinse down the mixer.

Water reducers, sometimes called plasticizers, are used to make concrete more workable with about 10 percent to 30 percent less water. Easy-to-work concrete on large jobs means lower labor costs; in addition, less water makes concrete stronger.

WORKING WITH CONCRETE

A concrete pour, even for small jobs, is a big event. You have only a certain amount of time before the concrete hardens, and you will be stuck with the results for a long time. But it needn't be a stressful event.

The key is to prepare well: draw accurate plans, gather tools, build forms, estimate exactly how much concrete you'll need, decide how you want the concrete delivered (bulk materials or ready-mix), arrange for delivery, get helpers, get inspections, and rehearse each stage of the pour. If you've taken time to do these tasks thoroughly, then working with the concrete itself—even if it involves many tons of material—will seem anticlimactic.

Note: Technically, concrete is *placed;* but because it is commonly referred to as being *poured,* both terms are used interchangeably in this book.

Planning the Job

Try to do all concrete projects at the same time, even if you can't finish the rest of any given project right away. It is easier to pour a front yard sidewalk and backyard deck footings when you have all the concrete tools, helpers, and materials assembled. You can finish building the deck at another time.

To plan the work, first calculate approximately how much concrete you will need, based on your working drawings and site measurements (see page 73). This does not have to be a precise calculation, which you will do later; just a rough estimate to help you make some preliminary decisions.

Should you do the concrete work yourself or hire a contractor? After reading this chapter, you will know what work is involved. Small jobs, of less than one yard of concrete, you can most likely do yourself. For large projects you have several options. You may want to do the excavating and then hire a concrete contractor to build the forms and complete the pour. Or, you

Concrete with an exposed aggregate finish is suitable for many types of patios, including informal (top) and formal (bottom) designs. The use of wood dividers in the patio at top makes it possible to work in small sections that can be poured and finished one at a time.

Tools for Concrete

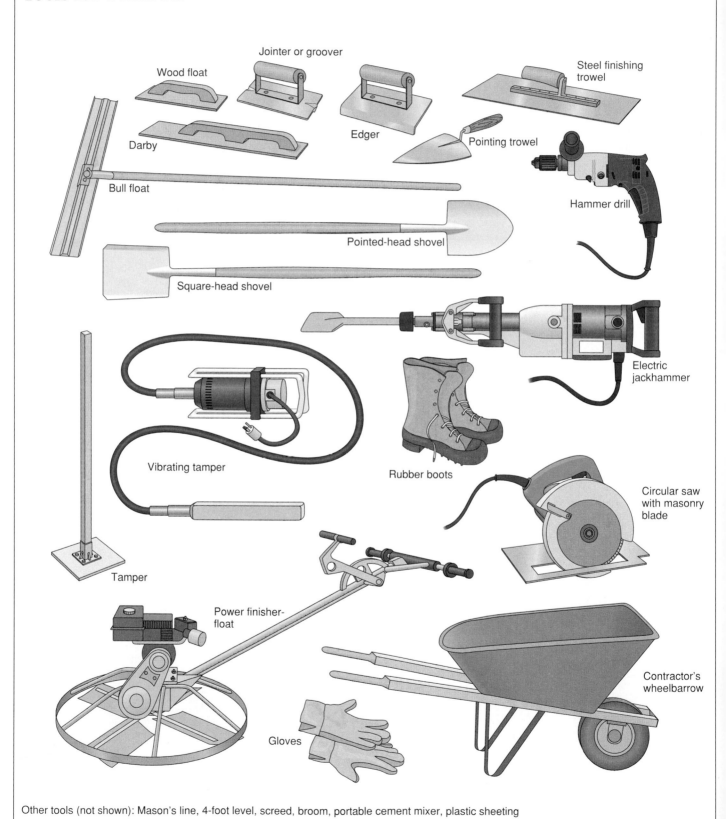

Wood float

Jointer or groover

Steel finishing trowel

Darby

Edger

Pointing trowel

Bull float

Hammer drill

Pointed-head shovel

Square-head shovel

Electric jackhammer

Vibrating tamper

Rubber boots

Circular saw with masonry blade

Tamper

Power finisher-float

Gloves

Contractor's wheelbarrow

Other tools (not shown): Mason's line, 4-foot level, screed, broom, portable cement mixer, plastic sheeting may want to do the excavations, build the forms, hire some moonlighting concrete finishers, and work as a member of the crew. Or, you may decide after building the forms that you're ready to do it all (with the help of some capable friends, if needed).

Another decision is whether to buy bulk materials and mix the concrete yourself, or to order a ready-mix delivery. As a rule, it is cost-effective to mix any amount up to ¼ yard yourself and to order a ready-mix delivery for any amount over 1 yard. For amounts between ¼ and 1 yard, you will have to weigh the higher cost of ready-mix delivery against the inconvenience and effort of mixing that much material (1 yard of concrete requires 564 pounds of cement, 1,400 pounds of sand, 2,000 pounds of gravel, and 272 pounds of water). See page 74 for more information about mixing.

Note: A cubic yard of concrete is usually referred to as a yard.

Another important part of planning is to get the proper tools (see illustration opposite). Some are inexpensive enough to buy, even for one project. Others you will have to borrow or rent, such as a rebar bender or cement mixer. Also, be sure to have several buckets (the 5-gallon plastic type work well), some straight 2×4s for screeds, and some short lengths of pipe or rebar for consolidating the concrete. Finally, have rubber gloves and, for large slabs, rubber boots for all crew members.

Building Forms

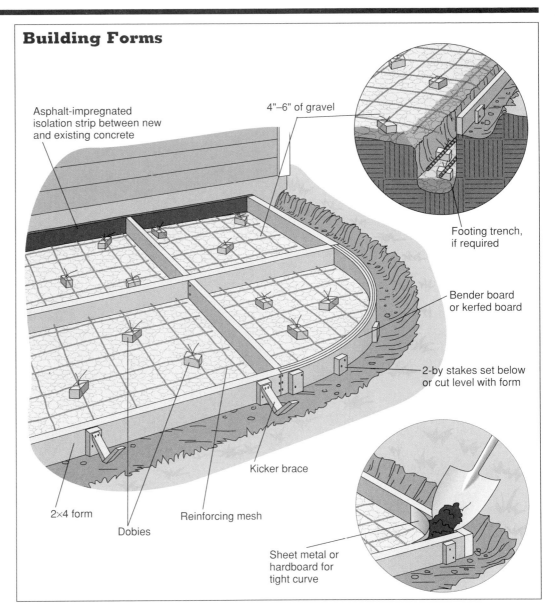

Asphalt-impregnated isolation strip between new and existing concrete

4"–6" of gravel

Footing trench, if required

Bender board or kerfed board

2-by stakes set below or cut level with form

Kicker brace

Sheet metal or hardboard for tight curve

2×4 form

Dobies

Reinforcing mesh

Forms and Reinforcement

Forms mold the concrete in a specific shape until it hardens. For some projects the forms are the ground itself—holes or trenches dug in precise dimensions for footings. For other projects forms must be built from lumber or plywood, or from prefabricated forming tubes. Specific forming techniques and dimensions for various projects are described later in this chapter (pages 80 to 93), but the following general techniques apply to most projects.

Excavating

When digging footing holes and trenches, keep the sides of the excavation straight, the bottom level, and the corners square. Use a square-nosed shovel. When using a clamshell digger or auger to dig round holes, keep the sides straight and flare the bottom of the hole slightly to enlarge the footing area. Do not overdig; you don't want to handle more concrete than necessary. Also, if you dig footing holes too deep, you cannot backfill the cavity with soil unless it is compacted to engineered specifications; footings must rest on undisturbed soil.

Building Forms

Align the inside edges of forms with the layout lines (see pages 14 to 16). Before nailing them to the stakes, make sure the tops of the forms are all even; measure down from your string lines or use a water level or other leveling device. Nail forms together and to stakes with duplex, or double-head, nails; they make it easy to adjust forms and to disassemble them after the concrete cures. Make sure that all joints are flush and cut off stakes flush with the tops of the forms to make screeding the concrete easier.

Any forms that are intended to become permanent edging or divider boards should be made from pressure-treated lumber or heart grades of a durable species of lumber, such as redwood. Drive 16d galvanized nails or 3-inch deck screws along the insides of the boards to lock them permanently to the concrete, and cover the tops with masking tape to protect them from becoming stained during concrete pouring and finishing,

Brace all forms securely. Low forms for flatwork (patios and walks), which are made from 2×4s, can be stabilized by stakes driven around the outside. Higher forms should also have "kickers," or diagonal braces, to secure the tops. Concrete is heavy. It develops a lateral, or outward, thrust of 150 pounds per square foot and can bow tall forms unless they are braced securely. Forms also have a tendency to float, or rise upward, unless they are secured to the ground by stakes or weights.

There are two ways to make curved forms. For a tight curve, where you are simply rounding off a corner, build forms as you would for

a rectangle. Then cut a strip of sheetmetal or thin hardboard, bend it into the corner, and nail the ends to the form boards. Pound stakes behind the strip or backfill it with gravel for support.

For larger curves, saw a series of kerfs from halfway to two-thirds of the way through a 2×4 so it will bend. Or, use strips of redwood benderboard or ¼-inch plywood and double or triple them for added strength (see page 71).

For patios and other flatwork, make a screed board from a straight 2×4 about 18 to 24 inches longer than the width of the surface of the concrete. If you are pouring a slab that is too wide for the screed board to span, set up temporary 2×4s to drag the screed across (see page 75). Support these screed guides with stakes so that their tops are level with the forms. Remove them after screeding and work concrete into the voids that they leave.

Coat the inside of the forms with form oil or old motor oil to keep them from sticking to the concrete. Avoid getting the oil on the ground or where you are pouring the concrete.

Where concrete will be butted up against an existing foundation or slab, set a fiber expansion strip or a strip of foam board between the old concrete and new pour. It will compensate for expansion and contraction of the concrete caused by any temperature changes, which might otherwise crack the foundation.

If you are placing new concrete against wood, such as siding boards, protect the wood by attaching aluminum

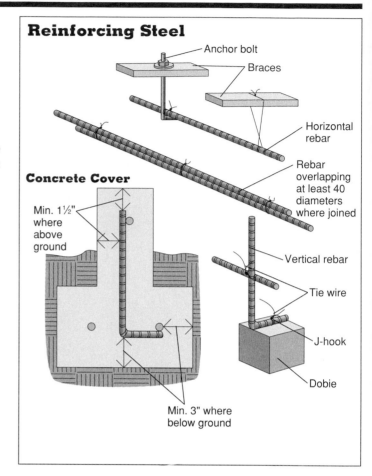

Reinforcing Steel

Anchor bolt

Braces

Horizontal rebar

Rebar overlapping at least 40 diameters where joined

Concrete Cover

Min. 1½" where above ground

Vertical rebar

Tie wire

J-hook

Dobie

Min. 3" where below ground

sheet metal to the siding where the concrete will be poured against it.

Placing Steel Reinforcement

Reinforcing steel, or rebar, commonly comes in 20-foot lengths and in diameters of ⅜ inch (#3), ½ inch (#4), and ⅝ inch (#5). It comes in two different strengths—40 grade and 60 grade. The 40 grade is adequate for all backyard projects. You can rent a tool for cutting and bending rebar, or you can cut it with a reciprocating saw and hacksaw blade, or a circular saw and metal cutoff blade. You can also improvise ways to bend rebar, such as having helpers stand on a board

placed over it while you slip a pipe over one end to the point where you want the bend and lift up on the pipe.

Rebar must be placed in the forms so that, when it is encased in concrete, it will be at least 3 inches away from any soil and, where the concrete is above ground, at least 1½ inches away from the air. These dimensions are referred to as concrete cover, and must be maintained to prevent the rebar from rusting and disintegrating over time. Three-inch concrete blocks, called dobies, are available on which to place rebar so it won't touch the ground. The dobies have wires to twist around the rebar. Do not use bricks for this purpose; they transfer moisture.

Where two pieces of rebar are spliced together end-to-end, the pieces must lap each other by at least 40 diameters (for example, 20 inches for ½-inch rebar). Use tie wire to secure splices and to tie rebars together wherever they intersect; wrap a short piece around both rebars and twist the ends of the wire together several times.

When setting pieces of rebar in place vertically, do not pound them into the ground; bend the bottom end into an L-shape and tie it to horizontal pieces of rebar so the bottom is at least 3 inches off the ground.

For flatwork, use concrete reinforcing mesh made of #10 wire in a grid of 6-by-6-inch squares (known as 6-6-10-10 wire). It comes in rolls 5, 6, or 7 feet wide. Cut the wire with bolt cutters and work with it carefully; it can coil up viciously unless both ends are secure. Use 2-inch dobies to hold it off the ground.

If your job requires a permit, have the excavation, forms, and rebar inspected before the concrete is placed. *Do not arrange for delivery until after the inspection.*

Estimating Volume

After the forms are built you can take precise measurements for calculating the amount of concrete needed. To do this, start with the simplest shapes, such as rectangular deck footings or a square patio. Multiply length by width by depth (always three dimensions), using feet and fractions of a foot, rather than inches, as the common

measurement. (Later you will convert cubic feet to cubic yards, the standard measurement for concrete.)

For example, if a patio is to be 10 feet long, 8 feet wide, and 4 inches thick, multiply $10 \times 8 \times \frac{1}{3}$, which equals 26.7 cubic feet. Since there are 27 cubic feet in one cubic yard, this patio would require approximately one yard of concrete.

For cylindrical shapes, figure the area of the circular cross section (pi $\times$ radius²) and multiply it by the total length.

Divide complex shapes into a series of simple shapes and calculate each one separately; then total them. For example, for an L-shaped patio, calculate the volume of each leg separately and add them together; for a T-shaped foundation wall, calculate the wide footing on the bottom as one continuous rectangle and the narrow wall as another.

Remember to convert all dimensions to feet and fractions of a foot before multiplying. When you have the total number of cubic feet for the job, convert it to cubic yards by dividing by 27. Add 10 percent to the total to account for irregular excavations. It is better to have too much concrete than too little. You can always have a small concrete job ready, such as forms for concrete stepping-stones, in case some concrete is left over.

Ordering Ready-Mix Concrete

Before calling a ready-mix company (listed under Concrete, Ready-Mixed in the Yellow Pages) for a delivery,

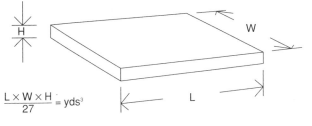

Calculating Volume in Cubic Yards*

Volume of a Cylinder

$$\frac{\pi \times R^2 \times H}{27} = yds^3$$

Volume of a Rectangular Solid

$$\frac{L \times W \times H}{27} = yds^3$$

*Note: All dimensions must be in feet or fractions of a foot

ask around at local masonry supply stores for recommendations. You may learn that some contractors favor certain companies for better service and consistent concrete.

When ordering, you will need to specify the amount of concrete, the delivery time and date, and type of mix. If you're not sure how to order the mix, simply explain what you are using it for. Also indicate any special conditions of the pour, such as abnormally hot or cold weather, and whether you are using a concrete pumping service. The following checklist will help you order.

• Amount of concrete, in yards. You pay for what you order, even if you don't need it all, so measure precisely.

• Delivery time. Morning is best, to allow enough time for finishing.

• Cement content, in sacks per yard. A 6-sack mix is desirable for most projects.

• Aggregate size. For most jobs, ¾-inch aggregate is suitable. If the concrete will be pumped, find out the maximum aggregate size for the size of hose being used. For some it is ⅜ inch.

• Water-to-cement ratio. Specify .5 or .55.

• Slump. This term is used to describe the consistency of fresh concrete. It refers to the number of inches a 12-inch high tower of fresh concrete slumps, as measured by a test involving a cone-shaped device. A 1-inch slump is very stiff, a 10-inch slump very soupy. For residential projects, 4 inches is average.

• Strength, or load-bearing capacity. Most residential

needs vary from a minimum of 2,000 pounds per square inch (psi) to over 4,000 psi. A 6-sack mix with a water/cement ratio of .5 should attain these strengths with proper handling and curing.

•Additives. Order air-entrainment for cold climates. It is specified as a percentage, usually 6 percent or 7 percent. You may also need an accelerator for cold weather, or a retarder for hot weather, but these additives are better left to the experts (see page 68).

•Cost. In addition to the per-yard charge, there will be a short-load fee for small orders (usually less than 4 yards). A stand-by charge is also customary: This is for any additional time the truck must remain at your site after a certain limit (usually about 5 minutes per yard). This fee can skyrocket if you are not prepared when the truck arrives. Clarify these fees ahead of time, and ask about any others.

Using a Pumper

If the site for placing concrete is out of reach of the delivery truck—chutes extend about 15 feet—you will need to order a pumper, either on your own or through the ready-mix company. Pumpers may cost $100 to $200 for a pour, which sounds expensive but may make the difference between doing the job or not. You don't want to risk a broken sidewalk or driveway, caused by a concrete truck backing over it. In addition, a pumper is faster than chuting from a truck and can cut labor costs.

Be sure that the pumper will arrive about a half hour

Mixing Concrete

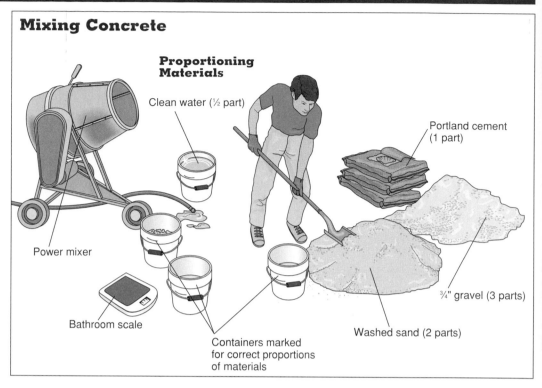

Proportioning Materials

Clean water (½ part)

Portland cement (1 part)

Power mixer

Bathroom scale

¾" gravel (3 parts)

Washed sand (2 parts)

Containers marked for correct proportions of materials

before the concrete is delivered, so there is time to set up. It takes at least two people to handle the hose after it is full of concrete—one to direct it and the other to support the hose as it is moved about the site. After the concrete is pumped, the pump truck driver will need to flush out the hose and hopper; have a wheelbarrow available and a place to dump the slop.

Mixing Concrete

For small jobs of less than 1 yard, concrete can be mixed on a sheet of plywood, in a wheelbarrow, or in a cement mixer, which you can rent. For very small amounts, buy bags of premixed ingredients. Most bags are 80 pounds, enough to make ⅔ cubic foot of concrete. This means 40 bags (3,200 pounds) per yard. For amounts over ½ yard, consider buying bulk ingredients (see chart).

Proportions of Ingredients for Making Concrete

	Cubic Feet of Concrete				
	4	6	12	18	27*
Cement (90-lb sacks)	1	1½	3	4½	6
Sand (pounds)	200	300	600	900	1,400
Gravel (pounds)	300	450	900	1,350	2,025
Water (gallons)	5	7½	15	22½	33¾
or Water (pounds)	40	60	120	180	270

*1 cubic yard

Be careful when mixing concrete. Follow all cautions printed on the bag, such as protecting your skin from wet cement. When mixing bulk ingredients, use a separate bucket for each item. Mix in batches of 1 to 3 cubic feet. For the first batch, set each bucket on a bathroom scale and fill it to the proper weight; then mark the fill line on the bucket so you don't have to weigh

ingredients each time. Mix the dry ingredients together first, then add the water—sparingly and just a little at a time. Adding too much water weakens concrete and causes problems such as spalling. When you plop a shovelful of properly mixed concrete on a piece of plywood, you can cut it with a trowel and it will hold its shape yet be soft enough to pour and form—kind of like risen dough.

Placing Concrete

Have plenty of help on hand for the pour. Have ready tools—such as hose, buckets, wheelbarrow, screed boards, floats, trowels, and edgers—and any anchor bolts or post anchors that must be embedded in the wet concrete. Have rubber gloves and boots available. Decide who will handle the pumper hose, who will spread the concrete, who will consolidate it, and who will screed. Wet down the trenches, gravel bed, and forms so they won't absorb moisture out of the concrete.

Fill the forms farthest from the truck first. Place the concrete level with the tops of the forms, using shovels to distribute it. Do not throw the concrete. Place it carefully. Otherwise, the aggregates might separate and weaken the mix. Consolidate, or settle, the concrete by jabbing a shovel or rod up and down in it. For large jobs, rent a vibrating tamper to speed the process. The concrete must be dense, but not packed, for maximum strength and durability.

For foundation walls, pour all of the footings first and let the concrete set up for a few minutes; then go back and fill the wall forms. Otherwise, the concrete in the trenches will be forced out by the massive weight of concrete in the tall forms. See pages 89 to 93 for more information about foundations.

Once the forms are filled, level the concrete with the top of the forms—a process known as screeding, or striking off. This is done by placing a straight 2×4 across the forms, then working it forward in a back-and-forth sawing motion. If there are humps or depressions in the concrete after your first pass, go over it again. Shovel more concrete into any low spots. Do a final consolidation of the concrete with a rod or a vibrating tamper and by hammering the outside of the forms to remove any air pockets. Remove any temporary screed guides (see page 72), and shovel new concrete into the cavities.

Floating, Edging, and Cutting Joints

The next step is a preliminary smoothing, called floating, which pushes aggregates below the surface. For small surfaces that you can reach easily, use a hand float or an elongated version called a darby, which extends your reach about 2 feet. For large slabs, use a bull float—a wide, flat board attached to a long handle. The board on all three types of float is made from either magnesium or wood. A magnesium float resists sticking better; it should always be used for air-entrained concrete.

To use a float, push it away from you with the leading edge slightly raised so that it will not dig in. Pull it back in the same manner, with the leading edge raised. Overlap each pass with the float until you have covered the entire pour.

Striking Off, or Screeding

Screed made from straight 2×4

Sawing motion

Edging or temporary screed guides

Floating Concrete

Bull float

Darby

Hand float

Edging and Cutting Joints

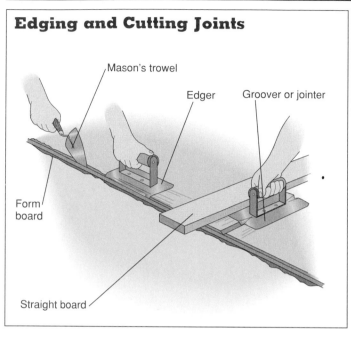

Mason's trowel

Edger

Groover or jointer

Form board

Straight board

Finishing Concrete

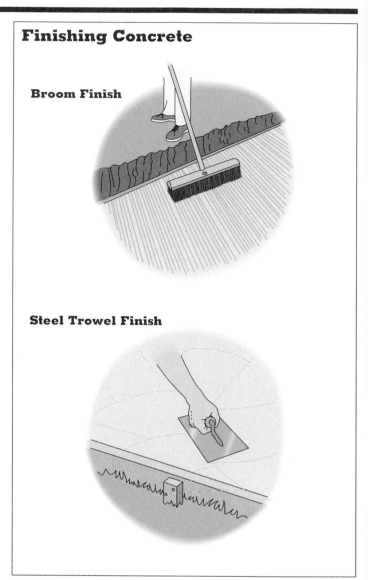

Broom Finish

Steel Trowel Finish

After floating, separate the concrete from the form boards by cutting along each form with a trowel. If the forms are to be removed, smooth and round the top edges of the concrete with an edging tool. This will not only improve the appearance, but the rounded edges will not readily crack or chip off.

At this time you should also cut control joints, or grooves, into the concrete. Concrete exposed to the weather swells and shrinks with the seasons, resulting in irregular cracks. Control joints make it more likely that cracks will occur underneath them, where they won't be seen. For patios, place control joints every 10 feet, in both directions. For walks, place control joints at intervals of 1½ times the width of the walk. If you used divider boards in the patio or walk, they will serve as control joints. Otherwise, cut the control joints with a jointing tool, using a straight board laid across the forms as a guide.

Control joints are not necessary in slabs that will be enclosed, such as garage floors.

Applying the Final Finish

The final finishing is done with various tools and techniques, depending on the texture you want. For a slightly rough texture, simply repeat the initial floating with a wood float. For a patterned nonslip surface, drag a dampened broom across the wet concrete in straight lines, curves, or a wavy pattern. For best results, use a soft broom designed for this purpose, and smooth the concrete with a steel trowel first. If you wish, you can run smooth borders and dividing strips around the broomed texture with the edging and grooving tools.

The slick surface found on basement and garage floors is called a steel trowel finish and is not ideal for outdoors, except as a base for brooming and salting. Steel-troweling a large slab is very difficult

for a beginner. It requires hand-troweling the surface two or three times, when the concrete is hard enough to support your weight on knee boards but fresh enough to produce a moistened paste as you smooth it. Unless you want to experiment on some small projects first, you should hire a skilled concrete finisher. In hot weather, when the concrete will harden faster than you can finish it by hand, it is worthwhile to rent a power finisher, known as a helicopter or whirlybird.

Important: Any kind of finishing must not be done until the sheen of water on the surface of the concrete has disappeared. This will take only minutes in hot, dry weather and an hour or more in damp, cool weather. If the concrete begins to set up before the sheen has disappeared, sweep off the water with a broom, soak it up with burlap, or drag the surface of the concrete with a hose, all without stepping on the concrete. Finishing done while water remains on the surface will result in dusting,

Custom Finishing

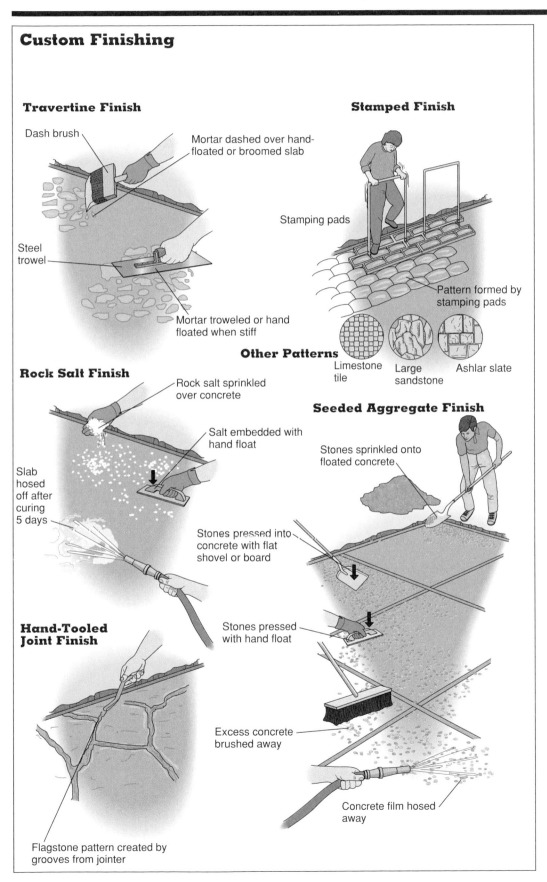

Travertine Finish

Dash brush

Mortar dashed over hand-floated or broomed slab

Steel trowel

Mortar troweled or hand floated when stiff

Stamped Finish

Stamping pads

Pattern formed by stamping pads

Other Patterns

Limestone tile

Large sandstone

Ashlar slate

Rock Salt Finish

Rock salt sprinkled over concrete

Salt embedded with hand float

Slab hosed off after curing 5 days

Hand-Tooled Joint Finish

Flagstone pattern created by grooves from jointer

Seeded Aggregate Finish

Stones sprinkled onto floated concrete

Stones pressed into concrete with flat shovel or board

Stones pressed with hand float

Excess concrete brushed away

Concrete film hosed away

---

spalling, and other problems after the concrete has cured.

Besides a float finish, broom finish, or steel trowel finish, there are several other decorative finishes you can do that require additional tools or materials. These include exposed or seeded aggregate, rock salt, travertine, stamped-pattern, and coloring.

Exposed Aggregate

You can expose aggregate that is already embedded in the concrete, or you can buy more attractive stones to "seed" the concrete surface. For seeded aggregate, divide the project into manageable segments so the concrete doesn't harden before you can work in the stones. You can divide the surface with boards, bricks, or tiles set in patterns. Set them about ½ inch lower than the forms. After pouring and screeding the concrete, sprinkle the aggregate across the surface, covering it evenly. With a helper, press the aggregate into the concrete with long 2×6s or flat shovels. Embed the stones firmly until you can just see their tops. Go over them with a wood float, if necessary, to push them down more. For unseeded aggregate, simply float the concrete and proceed to the next step.

When the concrete has hardened enough to support your weight on knee boards— usually three to five hours— sweep away the excess concrete around the stones with a stiff nylon brush or broom. Work carefully so that you do not dislodge the stones. Brush away the debris, spray a curing agent over the slab, and cover it with a plastic sheet for 24 hours. Then repeat the brooming process, this time with a fine spray of water, until about half of each stone is exposed. The water spray should be strong enough to wash away the concrete loosened by the broom but not disturb the surface of the concrete. Let the aggregate sit for another couple of hours, then hose off any concrete film on the stones. Cover the area with plastic and allow it to cure.

Rock Salt Finish

This is a slightly pitted, roughened surface created by sprinkling rock salt over concrete that has just been finish floated or steel troweled. Then lightly float the concrete again—just enough to press the rock salt flush with the surface. Cure it for five days under plastic sheeting, then wash and brush the surface. Don't use this technique in areas with freezing winters, because water freezing in the holes will damage the concrete.

Travertine Finish

This attractive surface resembles travertine marble. After the slab has been edged, apply a broom finish. Mix up a mortar of sand, water, white portland cement, and color pigment. Using a dash brush, spatter the mortar onto the slab. When the concrete can support you on knee boards, use a steel trowel to flatten the mortar and spread it, leaving voids in the low areas. This finish is not recommended for areas with freezing weather.

Stamped-Pattern Finish

A steel stamping tool, which can be rented, is pressed into fresh concrete to produce patterns resembling brick, cobblestone, flagstone, and others. Stampers also come in rubber mats and are used in the same way.

Before stamping concrete, it is best to measure the base of the stamp you plan to use and adjust your forms accordingly, so that the pattern comes out evenly. Two stamps should be used side by side. You place one, stand on it, then step onto

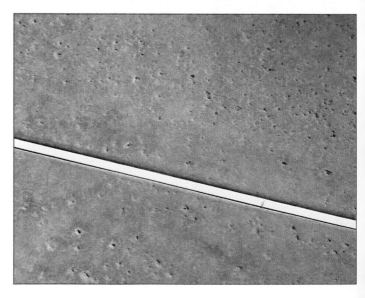

Concrete finishes: Above, stamped pattern with a border smoothed by steel trowel; top right, exposed aggregate; and lower right, rock salt finish.

the adjoining one. Impressions in the concrete should be about 1 inch deep. A jointing tool is then used to smooth out the pattern. Stamped concrete is usually colored.

Coloring

The best way to color concrete is to add the coloring agent, a pigment available from concrete suppliers, to the entire concrete mix. This way, the color is mixed throughout the concrete. For large jobs, however, this can be expensive. One alternative is to pour uncolored concrete to within 1 inch of the top of the form, then color the concrete mix used for the final layer. You can also color fresh concrete before it hardens by sprinkling a powdered coloring agent over it, hand-floating the dissolved powder into the surface, and repeating the process immediately. This is called the dry shake method.

Two ways to color hardened concrete are with chemical stains or paint. These should be applied only on concrete that is at least one year old. Both are specialized products; chemical stains should be applied by a licensed contractor. These finishes, especially paint, barely penetrate the concrete and may wear off.

Curing

Concrete must cure properly for it to be strong. Although concrete will harden in a day, it doesn't reach adequate strength for five to seven days. During this time, the concrete must be protected from drying out. To prevent evaporation,

cover the concrete with plastic (if the temperature is cool, use black plastic). Weight the edges and any seams with small stones or boards to trap as much moisture as possible. Let the concrete remain protected this way for about five to seven days. (Engineered projects where maximum strength is required, such as highways and bridges, require curing for at least 28 days.)

Another way to facilitate curing is to spray or roll on a curing agent. Clear or pigmented white, these agents keep the concrete moist. Do not use a curing agent if you plan to cover your slab with tile, bricks, stone, or flooring materials—the adhesive will not stick to the concrete.

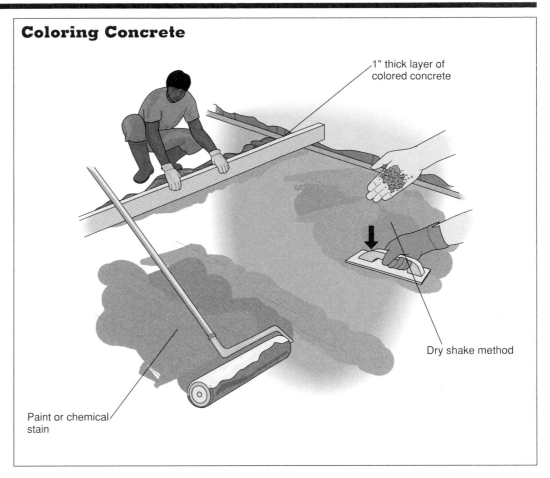

Coloring Concrete

1" thick layer of colored concrete

Dry shake method

Paint or chemical stain

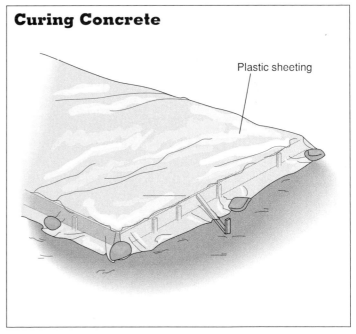

Curing Concrete

Plastic sheeting

WEEKEND PROJECTS

Many projects around the yard require at least some concrete. The projects covered here are not necessarily difficult—if you can build a small walk, you can build a complete patio or a basketball court. In addition to the techniques described for a specific project, review the basic techniques covered in the preceding pages.

Building a Patio

Building a concrete patio is relatively inexpensive and is within the abilities of most do-it-yourselfers. You can finish the concrete in interesting patterns and textures for the final surface, or you can use the concrete slab as a base for other paving materials.

Before planning your patio, contact the local building department to find out how thick a patio or walk must be, and how many inches of gravel should be placed under it to help prevent cracking when the ground freezes and thaws. See page 13 for more details about building codes and permits.

Laying Out a Patio

Refer to the first chapter for layout techniques. If the patio abuts your house or one side is parallel with a house wall, begin the layout along that wall and square the corners to it.

A patio should be sloped away from the house for proper drainage. To ensure proper slope, adjust the batter board heights as you do the layout. Rather than make all of the crosspieces level to each other, lower the batter board crosspieces at the outer edge of the patio at a rate of 1 inch for every 10 feet of distance from the house (see illustration, page 15). You can adjust the crosspiece height of the adjacent batter boards after stringing the lines.

After you have squared the perimeter string lines and adjusted them for slope, string new lines across the patio site every 4 to 5 feet to create a grid within the perimeter lines. Attach these grid lines to stakes or new batter boards; adjust the heights so that the grid lines just touch the perimeter lines where they intersect. This grid will make it easy to take height measurements as you excavate the site,

Excavating for a Patio

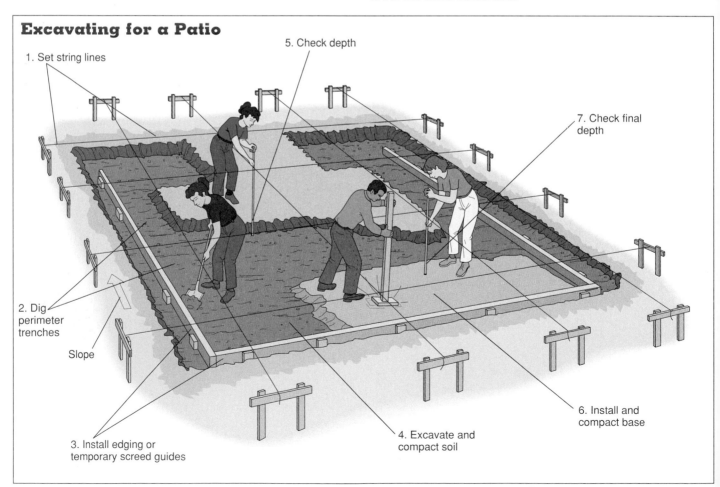

1. Set string lines
5. Check depth
7. Check final depth
2. Dig perimeter trenches
Slope
3. Install edging or temporary screed guides
4. Excavate and compact soil
6. Install and compact base

install the gravel base, and build the forms.

There is no rule that says that patios must have straight sides and right angles. If you are curving one of the sides, or building a freeform patio with no straight sides, establish a rectangular layout grid over the patio site, aligning perimeter lines over any straight sides of the patio. To lay out a curve, determine the center of the curve and drive a stake in the ground at that point. Tie a rope to the stake, stretch the rope taut, and tie a stick to it at the point where the curve begins. Swing an arc, scratching the ground with the stick to mark it. For freeform shapes, lay a hose on the ground, manipulate it into the shape you want, and pour sand or flour along it to mark the ground for excavating.

If you are simply rounding off the corners of a rectangle, do the layout as you would for a rectangular patio. You can build curves into the forms without laying them out ahead of time.

Excavating the Site

Plan the excavation so it is deep enough to accommodate a gravel base from 4 to 8 inches deep, plus whatever portion of the concrete slab must be below grade to bring the finished surface to the desired height—usually ground level. In areas where the ground freezes during winter, excavate a trench around the perimeter of the patio for a reinforced footing. Excavate to below the frost line, or to a depth required by the local building code.

Top: Striking off with a screed is the first step in leveling and finishing the surface of fresh concrete. Here two people are dragging a long screed board across the top of the forms in a back-and-forth sawing motion.
Bottom: This attractive walk has two different finishes—steel trowel border and exposed aggregate. Wooden strips, placed across the walk every 10 feet, absorb expansion and minimize cracking.

Before excavating, mark the patio dimensions on the ground with chalk or flour, then remove the layout strings. You will use them again later.

An easy way to begin excavating is to rent a rototiller to break up the sod. Or you can rent a power sod cutter or use a manual one. Then use a pick and shovel to excavate to the proper depth (invite friends to a digging party), or rent a small backhoe or dump loader. Start digging about 3 inches from the outside edge of the chalk line, to give yourself room to set the forms. Take frequent measurements from the grid lines so you don't overdig. Wear gloves so that your hands don't blister.

When the soil is removed, level the excavated area with a square-nosed shovel. Then moisten it and, using a roller or vibrating tamper, pack it down.

If you plan to install electrical outlets on the patio, now is the time to trench out the cable runs that you will later cover with concrete. Make sure to follow the local electrical code, and have all cable that will protrude through the concrete set in proper risers. All outlets should have ground fault circuit interrupter (GFCI) protection.

Installing a Gravel Base

The concrete should be poured on a bed of gravel to keep it stable, especially in areas with severe winters that cause frost heaves. Although you can add the gravel after building the forms, it is easier to install a gravel base before there are

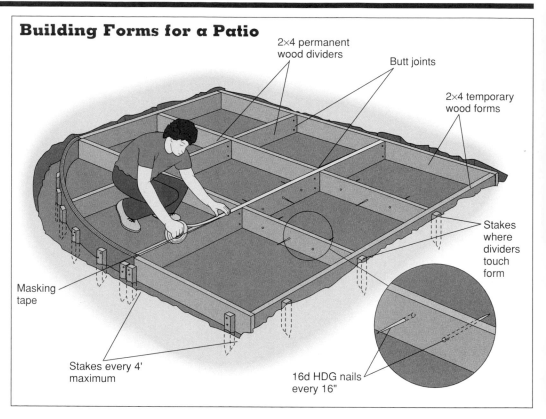

Building Forms for a Patio

2×4 permanent wood dividers

Butt joints

2×4 temporary wood forms

Stakes where dividers touch form

Masking tape

Stakes every 4' maximum

16d HDG nails every 16"

any forms in the way. Place crushed rock or pea gravel (class 5 gravel) on top of the soil base in 4-inch-deep layers. Spread it evenly with a rake, wet it down, and pack it with a vibrating compactor. Repeat this process until the gravel is the correct depth—4 inches below the finished patio surface—and feels solid.

Building the Forms

Build forms out of 2×4 lumber and stakes. The 2×4s are actually 3½ inches wide, so for a 4-inch-thick slab you will have to backfill some gravel or tack scraps of lumber along the bottom of the forms to prevent concrete from flowing under them. In areas with severe winters, which may require a thicker slab, use 2×6s or larger boards for the forms.

Align the inside edges of the forms directly under the string lines. Drive 1×2, 2×4, or steel stakes into the ground every 3 feet on the outside of the forms, and nail the stakes to the form boards with 8d (penny) or 16d (duplex) nails. If the forms are the permanent edging for the patio, use 16d hot-dipped galvanized (HDG) nails. Wherever two form boards meet, drive a support stake at the joint.

Slope the forms away from the house so the patio surface will drain properly. To do this, simply measure down from the layout lines an equal distance at all points. If the layout lines aren't sloped, use a transit or a carpenter's level and a long straightedge to determine slope while you're building the forms. Slope the boards away from the house at a rate of 1 inch per 10 feet

of distance. Also, don't forget to use an expansion strip against the house foundation (see page 72).

Although not usually required for patios (check your local code), it is advisable to put 6-inch reinforcing mesh in the slab. The mesh will help prevent the concrete from cracking due to shrinkage. Lay the mesh on 2-inch dobies so that it will be centered in the concrete. Also, place lengths of ⅜- or ½-inch rebar in any perimeter trenches, on 3-inch dobies (see page 72).

Installing Permanent Forms

For some patios the forms are permanent; they serve as a decorative wood edging. Many patios also have wood strips that divide the patio into sections. These serve as expansion joints and control

Pouring a Concrete Patio

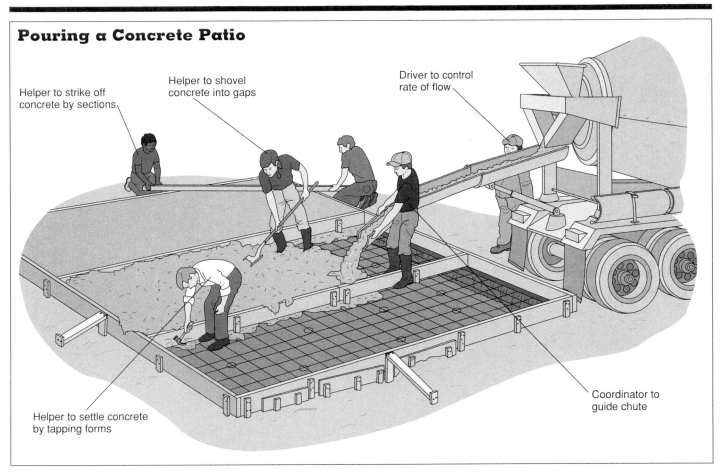

Helper to strike off concrete by sections

Helper to shovel concrete into gaps

Driver to control rate of flow

Coordinator to guide chute

Helper to settle concrete by tapping forms

joints, and should be placed no more than 10 feet apart.

Make permanent forms and divider strips with heart redwood, cedar, or cypress that has been given a coat of clear wood sealer, or with pressure-treated lumber, which requires no preparation except coating any cut ends with preservative. Join the boards at the corners with neat butt or miter joints. Drive the stakes 1 inch below the tops of the forms. Secure them by driving 16d HDG nails through the stakes into the boards. Also drive 16d HDG nails or galvanized deck screws horizontally into the boards, at midheight, every 16 inches where concrete will be poured against them. These help anchor the boards to the concrete. Cover the top edges

of the forms with masking tape to prevent stain marks from the concrete.

Providing Footings for an Overhead

If your patio design includes an overhead, also called a sunshade, arbor, or trellis, it requires posts with concrete footings. When you are building the patio forms, measure where each post will be and deepen the excavation an additional 6 inches to 12 inches at those points. Do not fill the holes with gravel. After you pour the concrete and it begins to set up, embed a post anchor or column base in the concrete for each post, making sure that it is properly aligned.

Pouring the Concrete

Before pouring the concrete, wet down the forms and gravel base so they do not absorb water from the concrete. Begin the pour in a corner at one end of the patio and fill the forms uniformly with concrete. Try to place concrete as close as possible to the final surface level. Work the concrete verti-cally—*not horizontally*—with a shovel or stake to break up any air pockets. Do not over-work the mix or you'll cause the aggregates to separate and come to the surface, weakening the mix. Use a vibrating tamper, which can be rented, and tamp just enough to settle the coarse aggregates below the surface.

As soon as the concrete has been spread and compacted, strike-off and float the slab. For a large patio, finish each section to this point before placing concrete in the next section. Do a complete section at once; avoid pouring new concrete directly against concrete that has already set up. This results in a cold joint, which is unsightly and tends to fracture easily.

After the concrete has set up enough for the watery sheen to be gone and for foot pressure to leave no more than a ¼-inch-deep indentation on the surface, proceed with the final finish and curing of the concrete. For details on how to do this, see pages 76 to 79.

83

Building a Concrete Walk

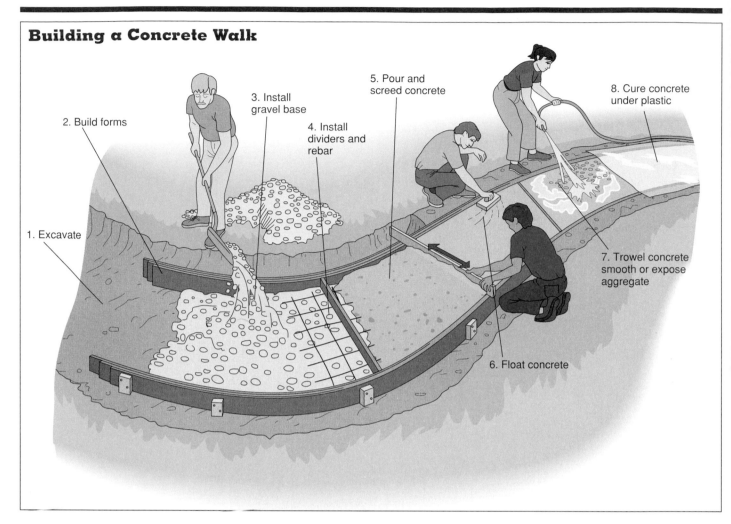

2. Build forms

3. Install gravel base

5. Pour and screed concrete

4. Install dividers and rebar

8. Cure concrete under plastic

1. Excavate

7. Trowel concrete smooth or expose aggregate

6. Float concrete

Building a Concrete Walk

A walk is usually easier to build than a patio because it is smaller in scale. The techniques for pouring and finishing the concrete are the same, but there are some differences in layout and form-building.

Laying Out the Walk

How wide should a walk be? Two people can walk comfortably together on a walk 4 feet wide; they can pass each other on one 3 feet wide. A person in a wheelchair needs a walk 40 inches wide, and a person

using a walker needs at least 27 inches.

To lay out the walk, stretch two parallel strings between stakes driven into the ground or, for a curved walk, position two hoses, in the desired shape. Mark the lines on the ground with chalk or flour, then remove the strings or hoses.

Excavate for the walk the same as you would for a patio (see page 81). Dig to the depth of the concrete, unless a gravel base is required due to unstable ground, or frost heaves where winters are severe. Plan for the surface of the walk to be slightly higher than ground level, so water won't puddle on the walk, but not so high

that you can't run a lawn-mower over it.

Building the Forms

Build forms out of 2×4s and stakes. On irregular ground, you can raise or lower the form boards to roughly follow the contour of the ground if it is not too steep. The most a walk should slope is 1 inch per foot.

If the site is steeper, build a stepped walk. The steps should all be of equal height, and equidistant from each other so that you do not have to adjust your stride as you walk on them. The easiest way to make a step is to place a

2×4 or 2×6 across the forms, then build the forms for the next section even with the top of the crosspiece. This crosspiece can be left in place if you want a decorative edging.

Pouring and Finishing the Concrete

Before pouring the concrete, wet the form boards. Fill the forms with concrete (see page 83) and screed it level with the tops of the form boards. Float the surface and cut control joints with a grooving tool (see page 76). They should be spaced at intervals of 1½ times the width of the walk.

Building a Backyard Basketball Court

If you can build a sidewalk or patio, you can build a backyard basketball court for shooting hoops with the kids.

You don't need a regulation high school court, which is 84 feet long by 50 feet wide. The minimum size of a court for two-on-two play is 20 feet by 20 feet. You may build it larger or smaller; the key to the length and width of the court is the 15 feet from the free throw line to the backboard. The hoop should be 10 feet off the ground.

You can buy a backboard mounted on a movable base, or you can permanently install an adjustable height pole. Both types are available at sporting goods stores. To help avoid injuries add padding to the pole.

Most pole manufacturers supply installation instructions. Basically, dig a hole about 2 feet deep by about 18 inches in diameter. Add a few inches of gravel to the hole for drainage.

Place the pole (make sure the top is capped so that water doesn't get in) in the center of the hole. Push stones against the pole to snug it up plumb as you fill the hole with concrete. With a level, plumb up the pole, using the stones in the hole as wedges. Let the concrete set and cure for five days.

Add the backboard and hoop to the pole after the concrete is cured. The backboard should overhang the court by about 15 inches to 20 inches.

At the same time you are setting the pole in concrete, you can also pour the court (think of it as a patio). When setting up the site, make sure to center the end of the court under the backboard. Now set up the forms, place a gravel base, and add reinforcing mesh to help keep the concrete from cracking. Using air-entrained concrete will also help. Finish it as smooth as you can.

To make the free throw lines, measure 15 feet in from the perimeter at each end and mark the location with chalk. To make 1-inch-wide lines, use masking tape on each side of the chalk line and paint in carefully with masonry paint.

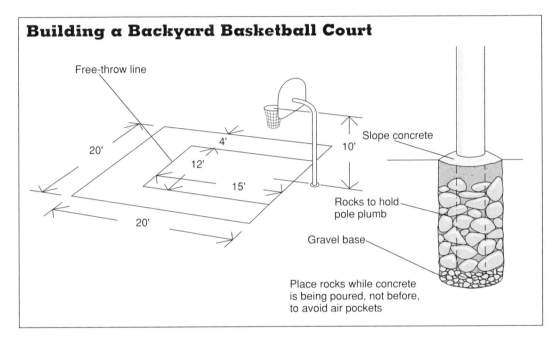

Building a Backyard Basketball Court

Free-throw line

20'

4'

12'

15'

20'

10'

Slope concrete

Rocks to hold pole plumb

Gravel base

Place rocks while concrete is being poured, not before, to avoid air pockets

Large slabs, such as this basketball court, require control joints or expansion strips every 10 feet to keep the concrete from cracking.

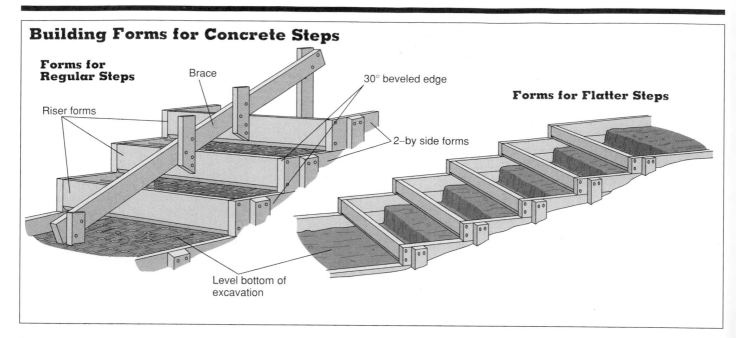

Building Forms for Concrete Steps

Forms for Regular Steps

Riser forms

Brace

30° beveled edge

Forms for Flatter Steps

2–by side forms

Level bottom of excavation

Building Steps

Concrete steps can be built on sloping ground to link different grade elevations of the yard. For gentle slopes, long steps with low risers are adequate. For steeper slopes, you will have to build steps with dimensions similar to conventional stairs. For either type, the finished surface can be the concrete itself, or it can be paved with brick, flagstone, or other materials. You must decide ahead of time which paving material you will use so you can include the thickness of the material in figuring the stair measurements.

Figuring Measurements

Building steps that are comfortable to walk on depends on using the correct rise (the height you raise your foot) and run (the tread, where your foot goes). Steps with treads that are too long or too short, or risers that are too tall or too short, are dangerous.

The rise and run of interior stairs are governed by building codes, which typically require that risers be no higher than 7½ inches and treads be about 11 inches deep. Exterior steps are usually more gradual and spacious. They can have about 12 to 15 inches of run for every 6 inches of rise. You also have more latitude in laying out an exterior stairway because the starting and ending points aren't fixed. For this reason, it is best to build the forms for the steps first, before building the rest of the walk, to ensure that the riser and tread dimensions are consistent throughout the stairway. The width of the stairway should be the same or wider than the walk, and should relate to the scale of the landscape.

Note: If you are building a set of steps between two fixed levels, such as a sidewalk and a patio, you will have to calculate the riser height so that the steps come out even—all steps must be the same height. To

measure, hold a long straight-edge perfectly level so that one end rests on the upper landing and the other end is suspended above the lower landing. Measure the distance between the bottom of the straightedge and the lower landing surface—this is the total rise. Divide this figure by a trial riser height, such as 6 inches; this yields the number of steps it takes to reach that height, which is most likely a whole number and a fraction. Round the number up to the

nearest whole number and divide this into the original total rise to get the precise riser height for each step. The tread depth depends on the riser height; the two added together should equal 18 inches to 22 inches.

Building Forms

The instructions that follow are for building forms for a casual set of steps that can be adjusted in total length to follow the general slope of the yard. If you are building

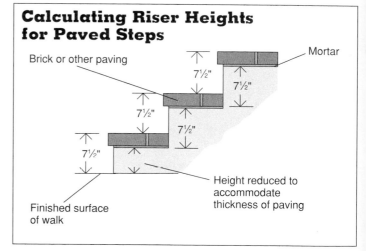

Calculating Riser Heights for Paved Steps

Brick or other paving

Mortar

7½"

7½"

7½"

7½"

7½"

Finished surface of walk

Height reduced to accommodate thickness of paving

steps that must begin and end at precise landing points, you will have to figure all measurements ahead of time (see opposite page) and adapt the following techniques to a specific layout.

To build forms, first lay two 2×12s on edge on the ground, following the slope, for the sides of the steps. Remove just enough soil to lower these stringers 3 inches or 4 inches into the ground, adjust them so that they are parallel and even with each other, and stake them in position.

Next, excavate the ground between the stringers into a series of stepped platforms. These do not have to conform to the finished concrete steps; they simply provide a level surface for the concrete to sit on. The excavation should be deep enough for the footing portion of the stairway, below the steps themselves, to be at least 6 inches thick—thicker if it must reach below the frost line.

To mark the stair layout on the inside of the forms, first mark a point at the bottom of one stringer that represents the finished surface of the walk leading up to the steps. Hold a level vertically at this point, so it is perfectly plumb, and scribe a line up from it. Mark this line at the chosen riser height, measured up from your first reference mark and, using the level, scribe a horizontal line at the upper mark. This line represents the first tread, which should be 11 inches long for 7-inch risers, from 12 inches to 15 inches long for 6-inch risers, and 16 inches for 5-inch risers. Repeat this process up the

stringer, using the same dimensions each time, until you reach the top of the grade. You may have to adjust the dimensions and start over again if the stairs are too flat or too steep the first time.

Important: If the concrete steps will be paved, *subtract* the thickness of the paving, plus bedding material, from the *first* riser height. All of the other risers should be the chosen riser height. Keep in mind that the finished surface of the top step will be higher than the layout tread line.

After laying out the first stringer, transfer the layout to the opposite stringer using a level and straightedge. For the riser forms, rip lengths of 2-by material to a width that is $\frac{3}{16}$ inch less than the riser height. Rip the boards at a 30-degree bevel so the bottom will be angled to allow space to slide a trowel under it while you finish the concrete on the step just below it. Cut these boards to length to fit between the stringers. Nail or screw them in place so the back edges align along your vertical layout lines and the tops align with the tread lines. (The bottoms are $\frac{3}{16}$ inch short to give the treads enough slope to prevent water from puddling on them.) Brace the boards with vertical cleats at each end, and, for steps wider than 30 inches, a long diagonal brace running the length of the stairway (see illustration, opposite page).

Alternately, you can build three-sided forms and stack them up the slope (see illustration, opposite page).

Pouring the Concrete

Pour gravel into the base of the forms and pack it down so that the gravel is flush with the bottom of the form. Add reinforcing mesh and rebar to help prevent cracking. Starting with the bottom step and working upward, pour the concrete so that it is flush with the tops of the horizontal form boards, then screed and trowel it. Let it cure for about five days.

Adding Paving

Start laying bricks (don't forget to wet them) or other paving on the bottom tread. Spread about a ½-inch-thick bed of mortar on the tread in an area just large enough to accommodate 3 or 4 bricks at a time. Allow a ½- to 1-inch

overhang on the front of the tread, and extend the bricks to the back of the tread. Any bricks you need to cut should be placed farthest from the edge of the step, butted against the riser.

Use a level to check that all the tread bricks are set at the same height, and use spacers to keep the width between all the bricks a uniform ½ inch.

Using the above technique, face the risers by placing the bricks or other paving units on edge (overlapping the back of the tread bricks), or simply leave the riser as plain concrete.

After laying all of the bricks or paving units, pack the joints with mortar or grout, then tool it smooth with a jointer.

Making custom stepping-stones out of concrete provides a way for you to exercise your creative whims.

Repairing Concrete

Despite the simplicity of making concrete, many things can go wrong in the process that will be in evidence years later. Here are repairs for some of the common problems.

Spalling

To patch small to medium-sized chips (spalls), use a masonry blade and a power masonry saw to saw the edges of the area, angling the cuts toward the undamaged concrete. Or you can use a chisel.

Chip out the concrete within the damaged area. Leave the edges rough and make sure that the remaining concrete is firm and sound. Vacuum out the dust, then fill the area with water. Let it soak into the concrete, then remove any excess water.

Prepare a mix of grout (1 part portland cement to 1 to 2 parts fine sand and enough water to make a creamy mix) and brush it over the area without filling the void completely. Mix a second batch, and let it stand for about 15 to 30 minutes to minimize shrinking. Smooth it over the first batch, compact it, and finish with a float or trowel. Let the patch cure for at least 3 days.

Broken Concrete

To mend small broken corners of steps or sidewalk slabs that are otherwise in good shape, clean the area, mix epoxy, and butter the broken piece. Hold it or brace it in place according to the manufacturer's directions. Scrape off the excess after it has set up.

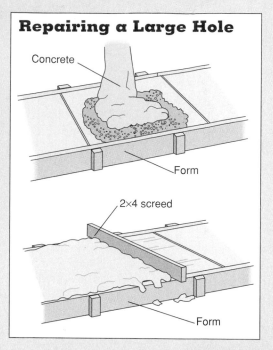

Repairing a Large Hole

Concrete

Form

2×4 screed

Form

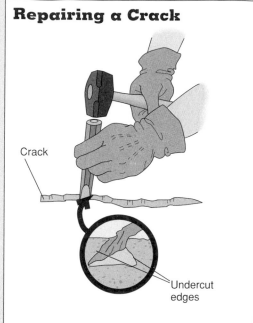

Repairing a Crack

Crack

Undercut edges

Corners with missing pieces can be repaired as if they were spalled. To help anchor the new concrete, install 6-inch lag screws with anchor shields. Drill 2 holes to a depth of 3 inches and use a socket wrench to set the screws.

Make a form and brace it in place. Wet the area; apply the grout and then the concrete. Use an edger to smooth the concrete against the form, then finish to match the rest of the surface.

Efflorescence

White soluble salts that come to the surface of concrete are called efflorescence. Although it is unsightly, it is harmless. Remove it with a wire brush, and flush the area with water. If that doesn't work, try a 5% to 10% solution of muriatic acid; dampen the area with water before application, and thoroughly flush afterward.

Whether caused by spalling or abrasion, broken concrete should be repaired as quickly as possible.

CONCRETE FOUNDATIONS

Concrete is an ideal material for foundations. It is strong and durable, able to support structural loads and to resist exposure to moisture. It can be placed anywhere and formed into a precise, level shape. It can be given a smooth finish, suitable for a floor surface. And it is economical.

Preliminary Considerations

While some foundations, especially on sloping ground or in areas with deep frost lines, are beyond the scope of this book, a simple garage or shed foundation is within the capabilities of most do-it-yourselfers. Be sure to check local codes and obtain any required building permits before you begin construction.

A slab foundation is simpler to form than a perimeter foundation, but it must be located on ground that is level or nearly so, and all rough electrical and plumbing must be completed before the concrete is poured. The finish must also be suitable for a floor surface, which may be the most difficult part of the project.

A perimeter foundation requires more extensive form building, but it can be built on irregular or sloping ground and the only finish required is striking off and floating the top of the wall.

Both types of foundation, no matter how well built, will only be as strong as the soil base they are built on. Footings must be deep enough to reach undisturbed soil—at least 12 inches below grade; deeper where the ground freezes—and must not be built on fill unless it is compacted to strict specifications. When building on uneven ground, have a bulldozer level the ground for you. The alternative for a sloped site is to build a stepped perimeter foundation. Short stud walls, called cripple walls, can be built on each step to produce a level support for the first floor.

Building a Slab Foundation

A slab foundation consists of a floor slab, which is 4 inches thick, and 12-inch-wide footings around the perimeter and under bearing walls. The depth of footings varies, depending on local codes (12 inches is minimum). Most techniques for building a slab foundation are the same as for a concrete patio (see pages 80 to 83). The following discussion presents the exceptions and variations to those techniques.

Constructing the Forms

Layout and excavation are the same as for a patio. Be sure that the outside of the excavation is aligned precisely, and not dug too wide, so the form boards can be set firmly on the ground along the edge.

Build 2-by perimeter forms high enough for the slab to be 8 inches above grade. Start placing the form boards at the highest corner and work toward the lowest corner. For a garage, slope the forms toward the door for drainage. The inside faces of the boards should be aligned directly under the string lines. Secure the boards with stakes every 18 inches. Fill any gaps under the form boards with scrap lumber nailed to the stakes and backfilled with soil.

Making Screed Guides

Because a garage is usually 20 or more feet wide, you may not be able to find a board long enough to use as a screed. There are two ways to solve this problem. One is to install permanent wood dividers in the slab, using pressure-treated or naturally rot-resistant 2×4s. The dividers should divide the

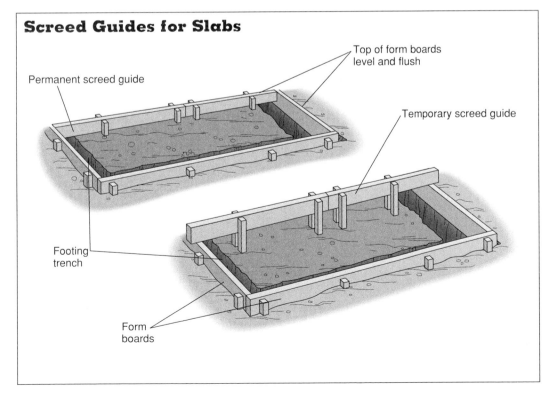

Screed Guides for Slabs

Permanent screed guide

Top of form boards level and flush

Temporary screed guide

Footing trench

Form boards

Slab Foundation Cross Section

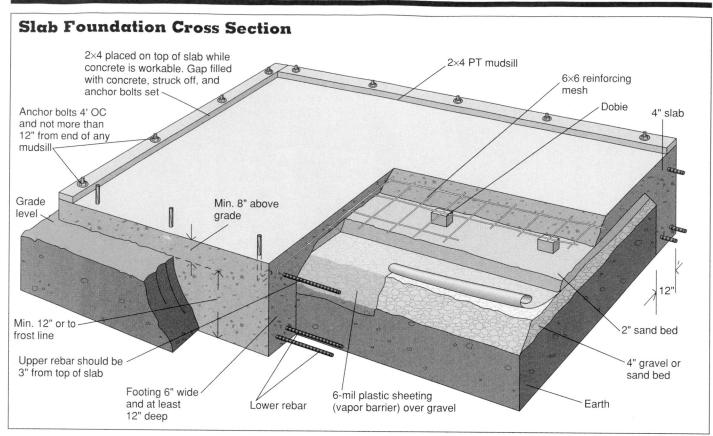

2×4 placed on top of slab while concrete is workable. Gap filled with concrete, struck off, and anchor bolts set

2×4 PT mudsill

6×6 reinforcing mesh

Dobie

4" slab

Anchor bolts 4' OC and not more than 12" from end of any mudsill

Grade level

Min. 8" above grade

12"

2" sand bed

Min. 12" or to frost line

4" gravel or sand bed

Upper rebar should be 3" from top of slab

Footing 6" wide and at least 12" deep

Lower rebar

6-mil plastic sheeting (vapor barrier) over gravel

Earth

---

slab into sections, no larger than 10 feet square. The tops of the dividers should be level with the tops of the forms.

The second type of screed guide is temporary, used to level the gravel base for the slab and to screed the concrete. Place two 2×4s, end to end, down the center of the slab. The outside ends should rest on top of the form boards; support the inside ends, where the two boards meet, with temporary stakes.

Make the screed itself from a 2×4. For temporary guides, nail an "ear" from a 1×2 stake to one end of the screed; the ear will ride on the temporary guide while the other end of the screed rides along the form boards. For more information about screeds, see page 75.

Placing Gravel, Vapor Barrier, and Steel

Before installing the gravel base, install any under-floor utilities, such as plumbing lines, and have them inspected. Then place a layer of gravel in the bottom of the excavation (but not in the footing trenches) and screed it level, 4 inches below the top of the forms (6 inches if you are adding a 2-inch layer of sand). Wet the gravel or sand and compact it with a plate vibrator. The gravel or sand should maintain a 45-degree "angle of repose" along the edge of the footing trench.

Next, place 6-mil plastic sheeting over the gravel or sand base, overlapping edges 12 inches, and 2 inches of sand over the plastic sheeting. Then

place horizontal rebar in the bottom of the footing trenches, on 3-inch dobies, and 6-inch reinforcing mesh throughout the slab area, on 2-inch dobies. (See page 72 for more information about reinforcing steel.)

Pouring and Finishing the Concrete

Before pouring the concrete, have the forms inspected if your project requires a permit. Coat the inside of the forms with form oil. Have the anchor bolts on hand, and mark locations along the form boards where they should be embedded in the fresh concrete. Have at least two helpers for the pour. Arrange for a pumper along with the ready-mix concrete delivery.

Fill the footing trenches with concrete first, then pour the slab, one section at a time. As work progresses, have helpers consolidate the concrete by jabbing a rod in the mix and by rapping the outside of the forms with a hammer. Using a garden rake, pull reinforcing mesh up into the center of the slab before you start screeding. If you have installed a temporary screed guide, you will not be able to work the screed back and forth with a sawing motion to consolidate the concrete. Use a vibrating tamper instead. Keep a close eye on the concrete as the pour proceeds. In hot weather, the first section of the slab may set while you are still pouring the remainder, so be ready to screed and float it as needed.

After screeding, remove the temporary screed guide and stakes. Fill the voids with concrete and smooth the slab with a bull float. After the concrete sets enough to place kneeling boards on it, begin finish floating the slab. Distribute your weight on kneeling boards made from 3-foot squares of ½-inch plywood, with 2×2s nailed along two edges for handles. Use the boards in pairs—one under your knees and the other under your toes.

Go over the surface once with a wood float for a coarse finish, and again with a steel trowel for a slick finish. Before the concrete hardens, cut between the forms and the concrete with a small triangular trowel.

Placing Anchor Bolts

Place anchor bolts as soon as the concrete has been screeded and bull floated. Place bolts in a straight line 1½ inches from the outside edge of the concrete. Embed the entire bolt except for the top 3 inches of threads. Bolts should be placed every 6 feet, on center (check the local code), and within 12 inches of each end of each mudsill. Avoid placing bolts in door openings.

Dealing With Interruptions

A ready-mix truck holds 8 to 9 yards of concrete. If you ordered more than that, the dispatcher will arrange a second truck to deliver the remaining concrete after the first. If you have to wait for a second truck to arrive, keep the concrete continuously

damp in the area where you stopped the pour. When you start pouring there again, work over the old and new concrete with a vibrating tamper to avoid a visible joint. Keep all of your tools and wheelbarrows washed and clean during an interruption.

Building a Perimeter Foundation

A poured concrete perimeter foundation consists of a continuous footing with a narrow stem wall on top, a configuration referred to as an inverted T. The footing and wall are usually poured at the same time—a monolithic pour. An alternative is a concrete block wall built atop a poured concrete footing.

Most codes require that the foundation for a single-story structure have a 12-inch-wide footing and a 6-inch-wide stem wall. For a two-story building, the footing should be 15 inches wide and the wall 8 inches wide. The footing should be as thick as the wall is wide, and at least 12 inches below grade (or below frost line). The top of the stem wall should be at least 8 inches above grade. Foundations higher than 3 feet should be built by professionals.

Layout and Excavation

Using layout techniques described on pages 14 to 16, set up string lines to mark the outside edges of the stem wall. Build the batter boards high enough so that the string lines will clear the top of the foun-

dation forms. All string lines must be level.

Using a plumb bob and tape measure, mark the ground with a continuous line of flour, sand, or gypsum for each side of the trench. The outside edge should be 3 inches outside the string lines for a 12-inch-wide footing, and 3½ inches for a 15-inch-wide footing.

Next, remove the layout lines and dig the footing trench to the required depth. Since concrete for the footing is poured directly into the footing trench, excavate carefully. Dig the sides straight, the bottom level, and the corners square. If the ground slopes, dig the bottom of the trench as a series of stepped, level platforms. Restring the layout lines to check your work—the stem wall must be centered over the footing.

Building Forms

Build the exterior side of the wall forms first, starting at the highest corner of the layout. First, drive several 3- or 4-foot-long stakes in a line, 3 to 4 feet apart, along the bottom of the footing trench. You can rent steel foundation stakes or buy 1×2 wood stakes. Steel stakes are easier to drive and, more important, to pull out of the fresh concrete before it sets up completely. The inside face of each stake should be exactly 1½ inches outside the string line to allow room for the board.

With a helper, hold the top form board in position; the top of the board should be at least 8 inches above the finished grade. Secure the board to the stakes with two 8d duplex nails at each stake. Complete the top course of the first

These foundation forms were built from lumber that can be reused as floor joists.

Building a Perimeter Foundation

Building Outside Forms

String line

Footing trench

Steel stake

Drive stakes into trench 1½" outside string line

Top form board almost touches string

Add form boards until they reach approximate top of footing

Duplex nails

Completing Forms

1×4 spacers, with anchor bolts, suspended in center holes

Hold steel tie straps with tapered pegs

Suspend rebar from forms with 16d nails

Pouring Concrete

Helpers poke and prod concrete to eliminate air pockets

First pass should fill footing trench, up to 3" above bottom of forms

Second pour should fill forms

THUMP THUMP

Rap forms to seat concrete against wood

wall. Add stakes, where necessary, to support the ends of boards. The last board can extend beyond the corner; you will butt the first board of the next wall against it.

After the boards are in place along the first wall, start the second course below them. Stagger the end joints so that they don't fall on the same stakes as the top course.

The number of courses depends on the height of the wall and the width of the boards. The bottom board should extend into the trench to a point just above the top of the footing. For a 6-inch-thick footing, the bottom of the lowest board will be about 6 inches above the soil.

Complete all four sides of the exterior wall in the same manner. Check frequently to make sure that the top boards are level by measuring down from the string lines; all measurements should be equal.

Bracing the Forms

Every 4 feet and at every corner, brace the forms with a 2×4 angle brace made from three 2×4s. The bottom of the brace should be 2 feet long, the upright leg about the same height as the wall, and the diagonal piece about 3 feet long. Cut each end of the diagonal piece at a 45-degree angle and nail the pieces together into a triangular shaped brace. Nail the brace to the forms with 16d duplex nails, align the forms with the string lines, and drive a short stake into the ground just behind the end of the brace. Secure the stake to the brace with two 16d duplex nails.

An alternative to building an angle brace is to drive a 2×4 stake diagonally into the ground. Align the form board with the string line, drive the stake, and nail it to the form board with two 16d duplex nails.

Hanging Steel

Steel reinforcing rods (rebar) must be suspended in the footing and in the foundation wall. Two horizontal rebars should be placed near the bottom of the footing, 3 inches from the soil; another rebar should be placed about 2 inches from the top of the foundation wall. If the wall is more than 18 inches high, a rebar should be hung between the top and bottom pieces.

Although you can place the rebar after both walls of the form are built, it is easier to install it before building the interior wall. Place the bottom rebar on 3-inch dobies. Hang the top (and middle) piece with tie wire secured to 16d galvanized nails driven about 3 inches below the top edge of the forms.

Building the Interior Wall

String new lines from the batter boards 6 inches (8 inches for two-story buildings) inside the original string lines. Drive foundation stakes 1½ inches inside that line, then begin hanging the form boards in the same manner as you did for the exterior wall.

After all of the boards for the interior wall are in place, tie the two walls together along the top with 1×4 braces, spaced 3 to 4 feet apart. Since they double as anchor bolt holders, make sure there is one where each anchor bolt will be placed.

Tie the lower courses of the wall together with form ties. These are metal braces that hold the walls exactly 6 (or 8) inches apart. They fit into the cracks between boards and will become embedded in the concrete. Each strap has a wedge at each end that locks the form board to it. Space the ties approximately 3 feet apart under the top course, and 18 to 24 inches apart for the lower courses.

Forming Openings in the Wall

To gain access to the crawl space under the building after the foundation has been poured and the floor built, you must create an opening in the wall. Basically a three-sided box, the form for the opening should be made from redwood or pressure-treated lumber. Use 2×6 material for a 6-inch-thick wall.

Hold the form in the wall so that the bottom of the form is just below ground level, then drive a few nails into the uprights to hold the form in place while the concrete is being poured. Work the concrete with a shovel as it is being poured so that it flows under the bottom of the form for the opening.

If girders will be used to support the floor joists, the inside of the concrete wall can be "keyed" to accept the ends of the girders. This is done by nailing a 2-inch length of girder lumber to the interior form precisely where the girder will fit.

If you are planning to run water, electricity, or heating ducts underground into the building, you will have to form holes in the footing.

This is easily done by placing a length of electrical conduit or plastic drainpipe across the footing trench before the pour is made. Dig a trench so that the pipe extends about 12 inches beyond the footing on each side and can be located readily after the pour is complete. Wrap the conduit or pipe with blanket insulation or roofing felt before the pour to cushion it from the expansion and contraction of the hardened concrete.

Pouring and Finishing the Concrete

Have the forms and steel inspected before you order and pour the concrete. Clear all debris from the trenches, oil the forms, and make sure that the rebar will be covered by at least 3 inches of concrete where it is below grade, and 1½ inches above grade. Drill ⅝-inch holes in cleats at the anchor bolt locations and hang bolts in them.

For the pour, have a minimum of two and preferably four helpers on hand. For specific techniques, see page 83. After the forms are filled and the concrete is consolidated, screed and float the top. After the concrete hardens enough to hold its shape, pull all of the stakes that became embedded in the footing concrete. The hardened concrete will keep the forms from collapsing. Then let the forms remain in place for at least 48 hours, preferably 72 hours.

INDEX

U.S./Metric Measure Conversion Chart

		Formulas for Exact Measures			Rounded Measures for Quick Reference		
	Symbol	When you know:	Multiply by:	To find:			
Mass	oz	ounces	28.35	grams	1 oz		= 30 g
(weight)	lb	pounds	0.45	kilograms	4 oz		= 115 g
	g	grams	0.035	ounces	8 oz		= 225 g
	kg	kilograms	2.2	pounds	16 oz	= 1 lb	= 450 g
					32 oz	= 2 lb	= 900 g
					36 oz	= 2¼ lb	= 1000 g (1 kg)
Volume	pt	pints	0.47	liters	1 c	= 8 oz	= 250 ml
	qt	quarts	0.95	liters	2 c (1 pt)	= 16 oz	= 500 ml
	gal	gallons	3.785	liters	4 c (1 qt)	= 32 oz	= 1 liter
	ml	milliliters	0.034	fluid ounces	4 qt (1 gal)	= 128 oz	= 3¾ liter
Length	in.	inches	2.54	centimeters	⅜ in.	= 1.0 cm	
	ft	feet	30.48	centimeters	1 in.	= 2.5 cm	
	yd	yards	0.9144	meters	2 in.	= 5.0 cm	
	mi	miles	1.609	kilometers	2½ in.	= 6.5 cm	
	km	kilometers	0.621	miles	12 in. (1 ft)	= 30.0 cm	
	m	meters	1.094	yards	1 yd	= 90.0 cm	
	cm	centimeters	0.39	inches	100 ft	= 30.0 m	
					1 mi	= 1.6 km	
Temperature	°F	Fahrenheit	⅝ (after subtracting 32)	Celsius	32° F	= 0° C	
	°C	Celsius	⅘ (then add 32)	Fahrenheit	68° F	= 20° C	
					212° F	= 100° C	
Area	in.²	square inches	6.452	square centimeters	1 in.²	= 6.5 cm²	
	ft²	square feet	929.0	square centimeters	1 ft²	= 930 cm²	
	yd²	square yards	8361.0	square centimeters	1 yd²	= 8360 cm²	
	a.	acres	0.4047	hectares	1 a.	= 4050 m²	